55th Annual Ontario Petroleum Institute Conference and Trade Show (OPI 2017)

Discover the Future

Grand Bend, Canada
24 – 26 May 2017

ISBN: 978-1-5108-4650-0

Printed from e-media with permission by:

Curran Associates, Inc.
57 Morehouse Lane
Red Hook, NY 12571

Some format issues inherent in the e-media version may also appear in this print version.

Printed by Curran Associates, Inc. (2017)

Ontario Petroleum Institute (OPI)
555 Southdale Road East
Suite 203
London, Ontario, Canada
N6E 1A2

Phone: 519-680-1620

opi@ontariopetroleuminstitute.com

Additional copies of this publication are available from:

Curran Associates, Inc.
57 Morehouse Lane
Red Hook, NY 12571 USA
Phone: 845-758-0400
Fax: 845-758-2633
Email: curran@proceedings.com
Web: www.proceedings.com

TABLE OF CONTENTS

HISTORY AND DEVELOPMENT OF OIL AND GAS PLAYS OF THE MICHIGAN BASIN

William B. Harrison, III

Michigan Geological Repository for Research and Education (MGRRE)

Western Michigan University-Michigan Geological Survey

1

EARLY HISTORY OF MICHIGAN OIL AND GAS EXPLORATION

- Some wildcatting as early as 1861, no significant finds
- First commercial wells near Port Huron in 1886-1887, 21 wells by 1910 producing up to 10 barrels per day
- Smith, 1912 reported over 100 wells that had been drilled in southeastern Michigan had natural gas and was being used for domestic heating, mostly from Antrim Shale or Traverse Limestone
- By 1920, five wells in the Deerfield Filed were producing oil from the Trenton Formation
- The Saginaw Field was discovered in the Berea Sandstone in 1925 and had 320 oil wells, 3 gas wells and 62 dry holes by the end of 1927
- Oil and Gas Permitting began in 1927
- The Muskegon Field in 1927 (Traverse Lime) and the Mt. Pleasant Field in 1928 (Dundee) established that oil production was possible throughout Michigan

2

Locations of Oil and Gas wells drilled throughout Michigan

57,441 boreholes drilled as of 3/28/2017

Red – Natural Gas
Green – Oil
Black – Dry hole

1,205 – Brine Disposal Wells
22,459 – Dry Holes
13,480 – Gas or Cond. Wells
3,026 – Gas Storage Wells
885 – Gas/other OBS Wells
106 – Type Unknown
59 – LPG Wells
506 – Mineral Wells
14,817 – Oil Wells
886 – Injection Wells

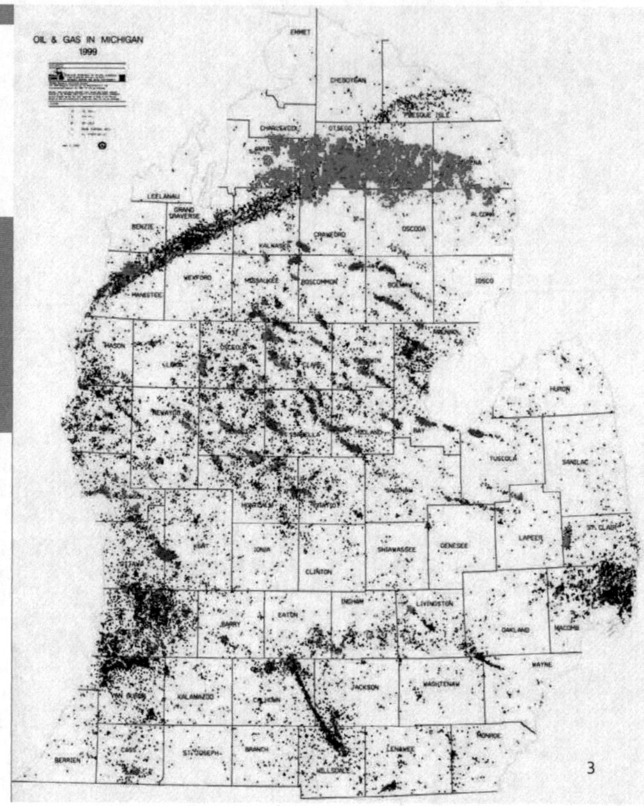

3

MICHIGAN OIL AND GAS ANNUAL PRODUCTION THROUGH DEC, 2016

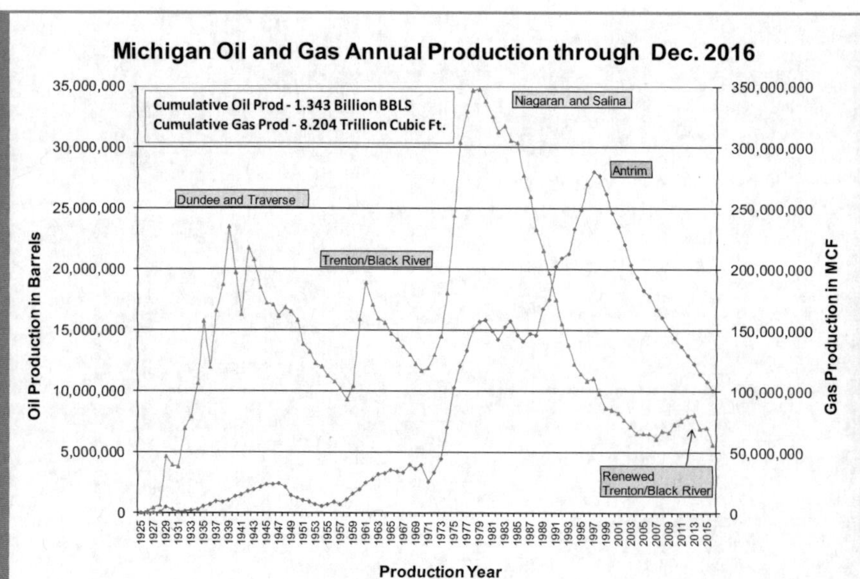

4

Major Carbonate Reservoir Intervals in Michigan Basin

From State of Michigan Geological Survey, 2000

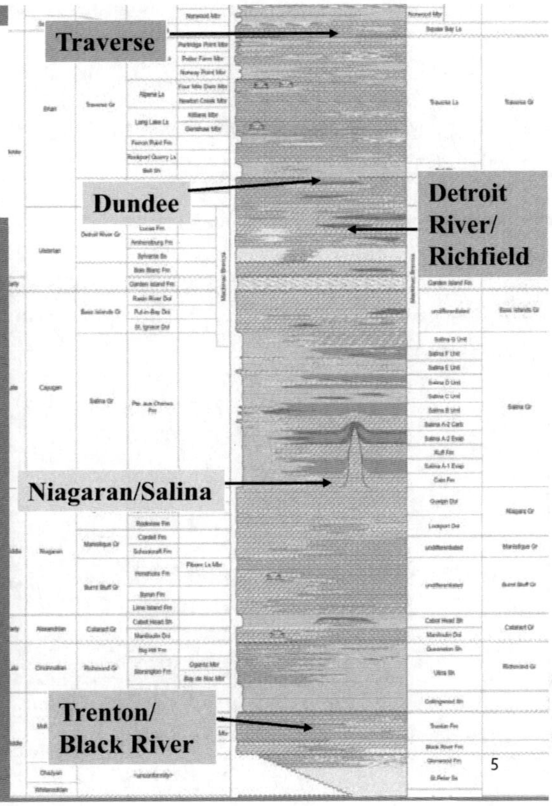

MAP OF OIL AND GAS WELLS FOR MICHIGAN'S LOWER PENINSULA CARBONATE RESERVOIR PLAYS

From State of Michigan Geological Survey, 2003

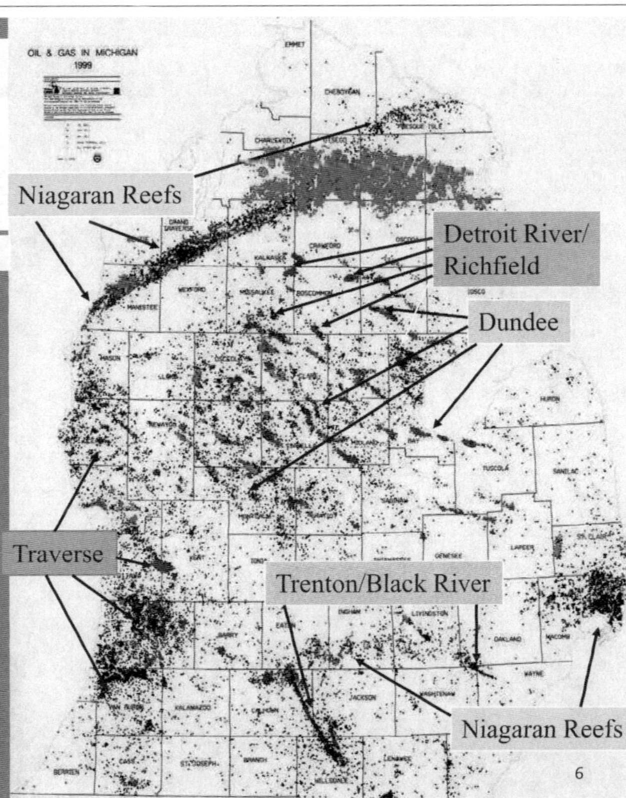

OIL AND GAS FIELDS PRODUCTION HISTORY - MICHIGAN CARBONATE RESERVOIRS

Formation	Number of Fields Reporting Production	Cumulative Oil Produced	Cumulative Gas Produced	Average Depth and Depth Range
Traverse	260	110 Million BBLS	13 BCF	2000 600 to 3400
Dundee	144	352 Million BBLS	42 BCF	3400 2200 to 4100
Detroit River/ Richfield	92	100 Million BBLS	120 BCF	4000 2000 to 5100
Niagaran	1187	450 Million BBLS	2500 BCF	5400 2300 to 7400
Trenton/Black River	19	140 Million BBLS	250 BCF	3900 2500 to 4700

7

MAJOR SILICLASTIC RESERVOIR INTERVALS IN MICHIGAN BASIN

From State of Michigan Geological Survey, 2000

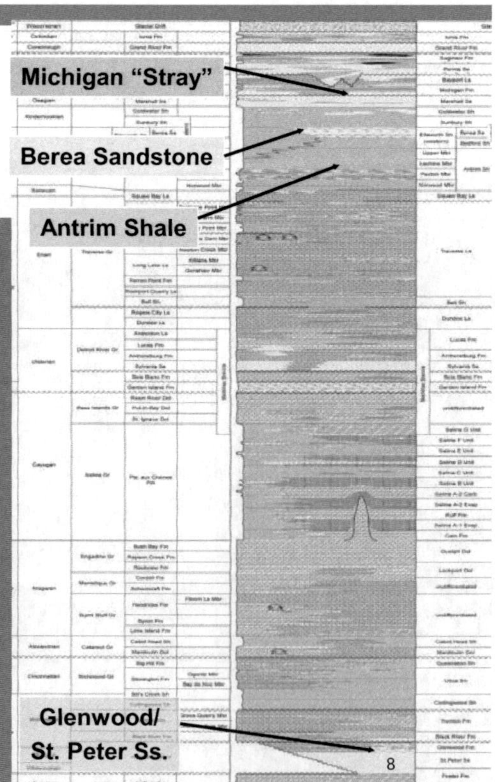

8

MAP OF OIL AND GAS WELLS FOR MICHIGAN'S LOWER PENINSULA SILICLASTIC RESERVOIR PLAYS

From State of Michigan Geological Survey, 2003

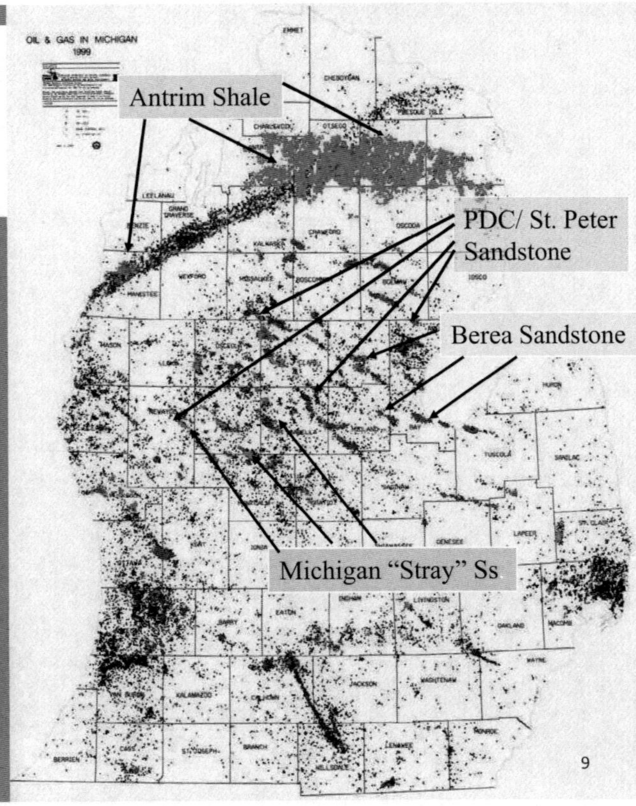

9

OIL AND GAS FIELDS PRODUCTION HISTORY MICHIGAN CLASTIC RESERVOIRS

Formation	Number of Fields Reporting Production	Cumulative Oil Produced	Cumulative Gas Produced	Average Depth and Depth Range
Michigan "Stray"	82	5 Million BBLS	215 BCF	1200 900 to 1800
Berea Ss.	53	8 Million BBLS	16 BCF	1500 800 to 2400
Antrim Shale	36	none	3335 BCF	1300 500 to 2600
St. Peter/ PDC	70	15 Million BBLS	600 BCF	9500 7000 to 12000

10

USGS PETROLEUM SYSTEMS CLASSIFICATION

- Based on source rock formation
- May have multiple reservoir formations for same source rock
- May have multiple seals for the different reservoir units
- TPS 1 – PreCambrian Nonesuch – no commercial production
- TPS 2 – Ordovician Foster
- TPS 3 – Ordovician to Devonian Composite
- TPS 4 – Silurian Niagaran/Salina
- TPS 5 – Devonian Antrim Shale
- TPS 6 – Pennsylvanian Saginaw – trivial commercial production

11

TPS 2 – ORDOVICIAN FOSTER

- Ordovician Sandstones and Carbonates AU
- Source – Foster Formation
- Reservoir – St. Peter Ss. (aka. PDC) and thin sandstone beds within the upper Foster Fm.
- Seal – Shales in the Glenwood Fm., Tight Limestones of Black River Group
- Trap – Structural anticlines on basement structures

12

FOSTER FORMATION – BRAZOS-ST. FOSTER CORE – OGEMAW COUNTY

A) Black laminated mudstone with isolated burrows (arrows)11,959 feet.

B) Laminated mudstone (dark) intercalated with laminar anhydrite12,398 feet

C) Laminated dolomicrite (dark) with intercalated laminar anhydrite (light). Localized very fine sandstone with ripple cross laminae (arrows)12,544 feet.

D) Laminated dolomicrite, cyanabacterial mats and anhydrite. Nodules of displacive anhydrite (arrow).11,731 feet.

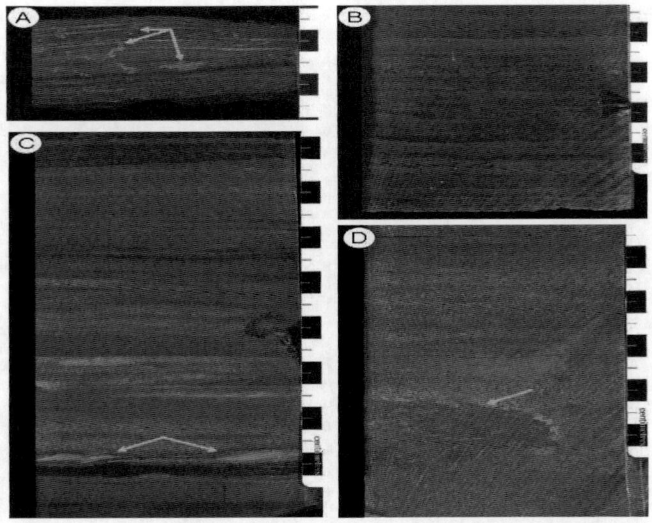

13

ST. PETER (AKA PRAIRIE DU CHIEN) SANDSTONE - RESERVOIR

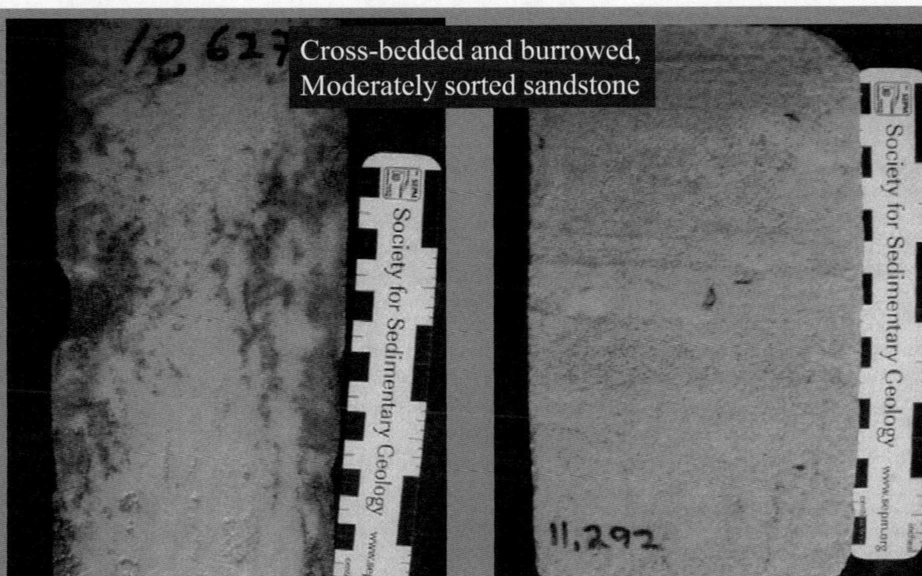

14

TPS 3 – ORDOVICIAN TO DEVONIAN COMPOSITE SYSTEM

- Ordovician Trenton/Black River AU
- Ordovician Collingwood Shale Gas AU
- Silurian Burnt Bluff AU
- Middle Devonian Carbonates AU
- Devonian Antrim Continuous Oil AU
- Devonian to Mississippian Berea/Michigan Sandstone AU

15

TPS 3 – ORDOVICIAN TO DEVONIAN COMPOSITE - ORDOVICIAN TRENTON/BLACK RIVER AU

- Source – Shale beds within the Trenton/Black River, Collingwood Fm., Utica Shale
- Reservoir – Fractured, vuggy and intercrystalline dolomite in T-BR
- Seal – Utica Shale or tight limestone of the Trenton/Black River
- Trap – Stratigraphic occurrence of porous Dolomite surrounded by tight Limestone

16

COLLINGWOOD FM. SOURCE ROCK AND UTICA SHALE SOURCE ROCK &SEAL

JEM- Bruggers #3-7 Core
Missaukee Co., MI

Petoskey Exploration – State Pioneer #1-3 Missaukee County

Utica Shale

Collingwood

17

TRENTON-BLACK RIVER RESERVOIR

18

TPS 3 – ORDOVICIAN TO DEVONIAN COMPOSITE - MIDDLE DEVONIAN CARBONATES AU

- Source – Amherstburg and Lucas Fms.
- Reservoir – Open marine shelf carbonates of the Traverse, Dundee and Amherstburg Limestones, Fractured and vuggy dolomite in the Traverse and Dundee and restricted carbonate facies in the Detroit River Group
- Seal – Bell Shale, Squaw Bay Fm., Antrim Sh., Tight Limestone, Anhydrite and Halite in the Lucas and Anhydrite in the Reed City
- Trap - Structural anticlines on basement structures, Porous Limestone from original depositional facies, porous dolomite from hydrothermal alteration, sucrosic dolomite from evaporative reflux

19

HYDROCARBON PRODUCING MIDDLE DEVONIAN ROCKS IN MICHIGAN

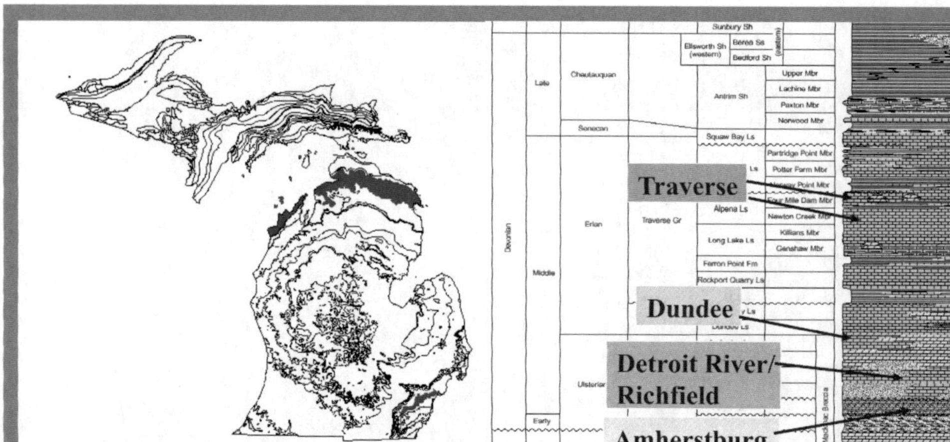

Middle Devonian
Traverse Lime, Dundee Fm., Detroit River Group and Amherstburg Fm.
Outcrop and Subcrop of these rocks shown in blue.

From Michigan Geological Survey Presentation – "The Rock Cycle"

20

AMHERSTBURG FM. – SOURCE ROCK AND RESERVOIR

Core Energy-St Charlton #4-30, Otsego Co.

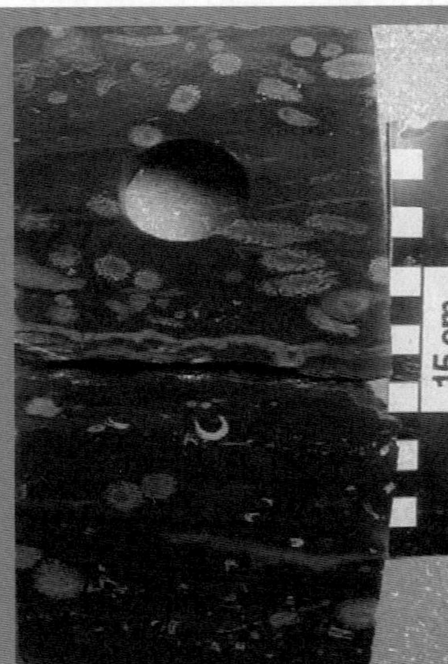

Hunt-McGuire #1-22, Oscoda Co.

21

TRAVERSE LIMESTONE RESERVOIRS – FRACTURED AND VUGGY DOLOMITE

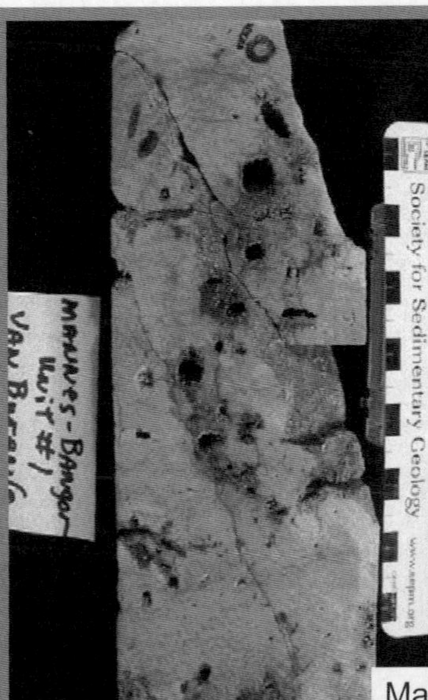

Mannes-Bangor Unit #1, Van Buren County

22

DUNDEE RESERVOIRS – LIMESTONE OR DOLOMITE

MichCon-LoReed #LR 83-2 Osceola Co. Wiser-Sturm #4-0 Gladwin Co. Cronus Dev. #1-3 Montcalm Co.

Laminated and Fenestral Wackestone and Mudstone | Reef and Reef Skeletal Debris | Skeletal Grainstone

23

BELL SHALE SEAL

Dart-Hamming #1-22, Missaukee Co._3886-3892 feet

24

LUCAS FM. RESERVOIR AND SEAL

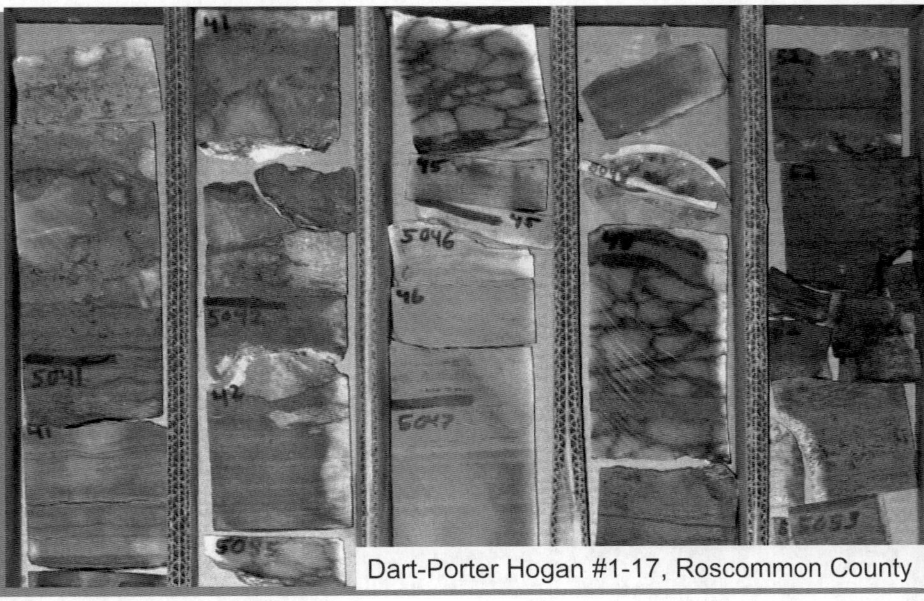

Dart-Porter Hogan #1-17, Roscommon County

25

TPS 3 – ORDOVICIAN TO DEVONIAN COMPOSITE - DEVONIAN TO MISSISSIPPIAN BEREA/MICHIGAN SANDSTONE AU

- Source – Antrim, Ellsworth and Sunbury Shales
- Reservoir – Berea Ss., Marshall Ss., Michigan "stray" Ss.
- Seal – Sunbury and Coldwater Shale, shales and anhydrite in the Michigan Fm.
- Trap - Structural anticlines on basement structures – mostly primary intergranular porosity, with minor cementation

26

MICHIGAN "STRAY" AND BEREA SANDSTONE RESERVOIRS

MichCon SL-175-A Six Lakes Field, Montcalm, Co | Wood-Mieske #1-6 Williams Field, Bay, Co

Michigan "stray" Ss.

Berea Ss.

27

TPS 4 – SILURIAN NIAGARAN/SALINA

- Silurian Niagara AU
- Silurian A-1and A-2 Carbonate AU
- Devonian Sylvania Sandstone AU
- No commercial production has been established in the Devonian Sylvania Ss. AU, so it was not quantitatively assessed
- Source – Salina A-1 Carbonate
- Reservoir – mostly Dolomitized Niagaran pinnacle reefs and associated facies
- Seal – Salina A-1 and A-2 Anhydrite and Halite
- Trap – Stratigraphic trap of carbonate buildups surrounded by draping evaporites. Most reservoir quality is in Dolomitized facies, minor porosity in Limestone

28

NIAGARAN REEF RESERVOIR

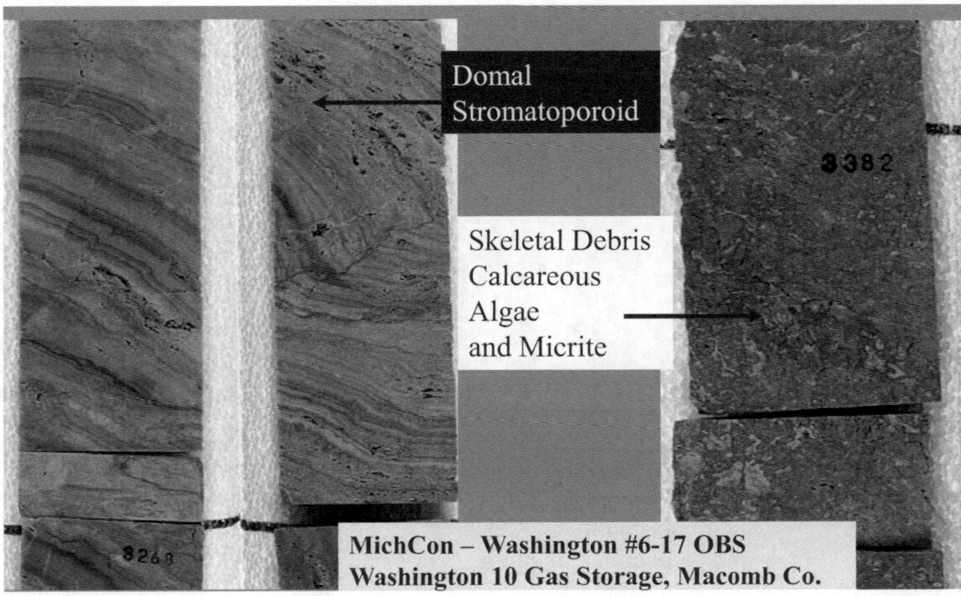

MichCon – Washington #6-17 OBS
Washington 10 Gas Storage, Macomb Co.

29

SALINA A-1 CARBONATE SOURCE/RESERVOIR

Organic-rich "poker-chip" facies Source rock

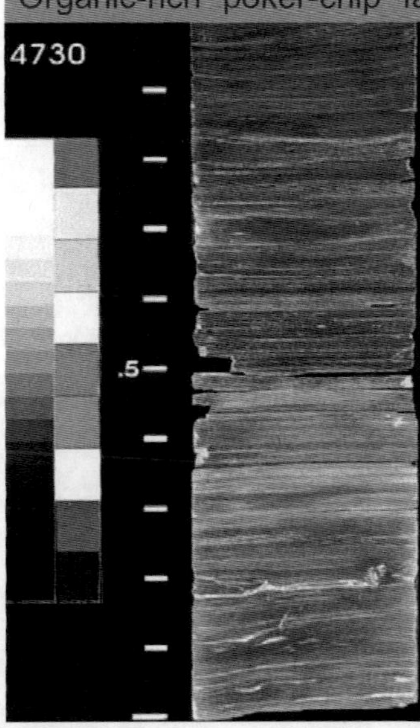

Rosetta Resources-Christensen #1-21P, Ionia County

Fractured and vuggy potential Reservoir

30

SALINA A-1 AND A-2 EVAPORITE SEAL

Chapman
Blair Brydges
Newaygo Co.
A-1 Salt
above A-0
Carbonate

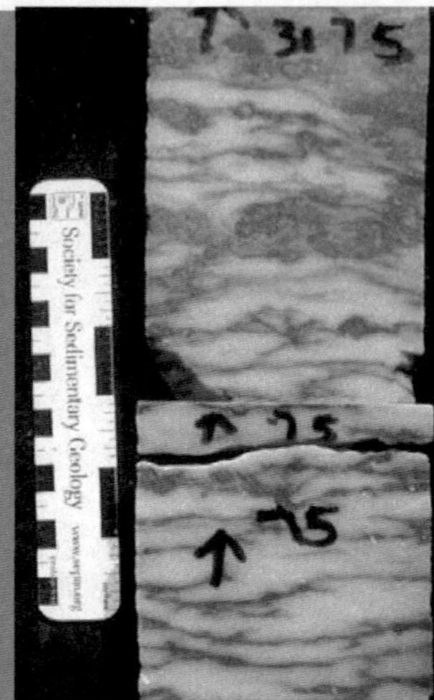

AMOCO-Miller-Fox #1-11
Oceana Co.
Salina A-2 Anhydrite

31

TPS 5 – DEVONIAN ANTRIM SHALE

- Devonian Antrim Shale Continuous Gas AU
- Source – Antrim Shale
- Reservoir – Antrim Shale
- Seal – Antrim Shale and Ellsworth laterally and vertically
- One of the first Shale gas producing formation in the U.S.
- Trap – Hydrodynamic, Gas is adsorbed to rock and desorbs into fractures as reservoir pressure is lowered. Reservoir has leaked into the glacial drift where it subcrops and charges some of the unconsolidated sands.

32

ANTRIM SHALE RESERVOIR AND SOURCE

Amoco-Conn D-2, Shiawassee County

33

CONCLUSIONS

- Michigan Basin Hydrocarbon exploration and production extends back to the 19th century
- Serious commercial production began in the 1920's
- Over 2200 fields are known from over 57,000 wells
- Most production is from carbonate reservoirs
- Total Michigan production is over 1.34 billion bbls oil and 8.0 TCF gas
- Michigan also leads the U.S. in underground Natural Gas storage, both working gas and total gas volumes
- Michigan reservoirs can be classified by the Total Petroleum System of source, reservoir and seal (U.S. Geological Survey)

34

MIDDLE DEVONIAN PLAY
MICHIGAN BASIN OF ONTARIO

Duncan Hamilton

PLAY BY PLAY
Exploring the Michigan Basin

1

Middle Devonian Pools of the Michigan Basin

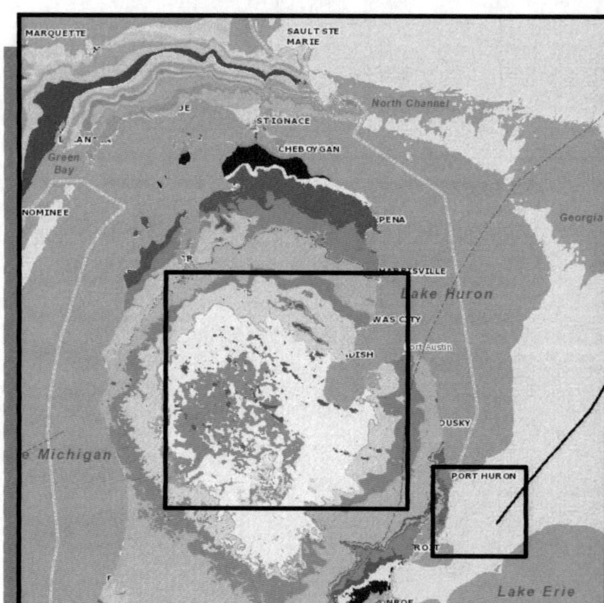

- 375+ Million Barrels Michigan
- 45+ million Barrels Ontario

PLAY BY PLAY
EXPLORING THE MICHIGAN BASIN

2

Outline of Presentation

- ➢ History of the Play
- ➢ Significant Devonian Pools
- ➢ General Play Parameters
- ➢ Field Example Rodney Pool
- ➢ Use of Modern Technology
- ➢ Future Potential

PLAY BY PLAY
EXPLORING THE MICHIGAN BASIN

3

History of the Play

Oil Springs Field, Lambton County – the site of the first commercial oil well in North America

- ➢ First commercial oil discovery at Oil Springs 1858
- ➢ 73% of oil discovered prior to 1900
- ➢ Last major discovery in 1949
- ➢ Initial flow rates as high as 7,500 bbls/d
- ➢ Produced 45+ million barrels to date
- ➢ Some of the original pools still producing
- ➢ Pools found mainly from oil seeps, tar beds
- ➢ Limited and sporadic drilling in recent times

PLAY BY PLAY
EXPLORING THE MICHIGAN BASIN

4

Significant Devonian Pools of Ontario

POOL NAME	DISCOVERY YEAR	CUMMULATIVE PRODUCTION (barrels)	ACTIVE
Oil Springs	1858	10,437,000	Yes
Petrolia	1862	18,798,000	Yes
Bothwell-Thamesville	1862	3,354,000	Yes
Wallacetown	1898	253,000	No
Glencoe	1917	1,139,000	No
Watford-Kerwood	1938	132,000	No
Rodney	1949	10,845,000	Yes

PLAY BY PLAY
EXPLORING THE MICHIGAN BASIN
5

General Play Parameters

- Stratigraphy
- Devonian Play Fairway
- Location of Pools
- Dundee Structure
- Structural Trap Formation Concept
- Summary of Play Parameters

PLAY BY PLAY
EXPLORING THE MICHIGAN BASIN
6

Stratigraphic Column

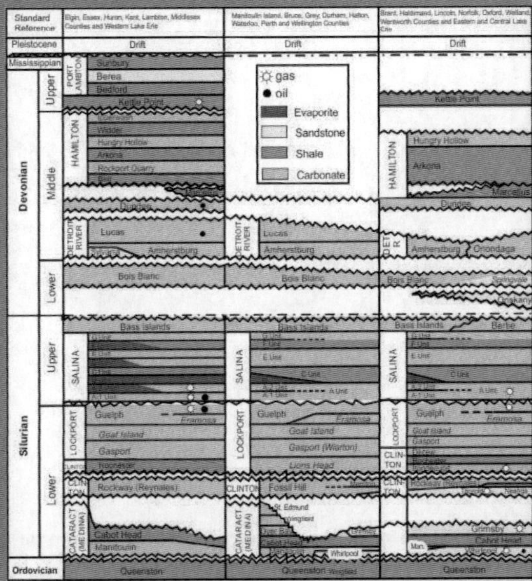

PLAY BY PLAY
EXPLORING THE MICHIGAN BASIN
7

Middle Devonian Play Stratigraphy

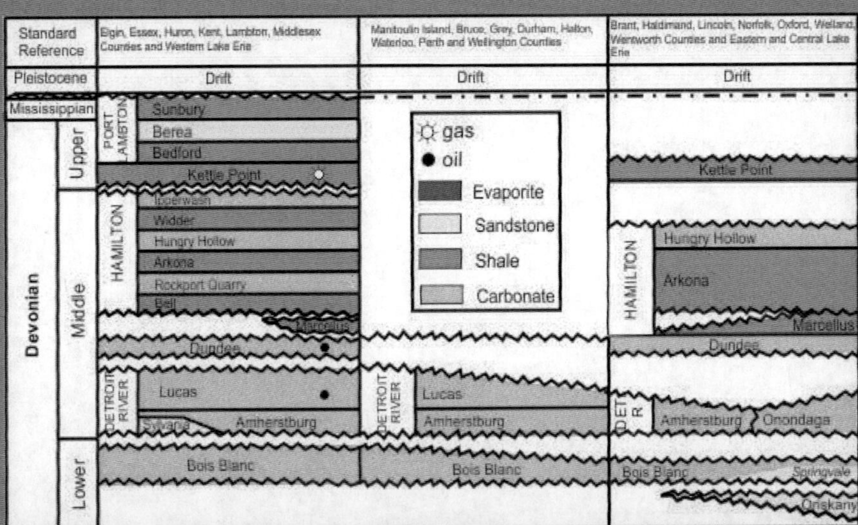

PLAY BY PLAY
EXPLORING THE MICHIGAN BASIN
8

Middle Devonian Pool Fairway

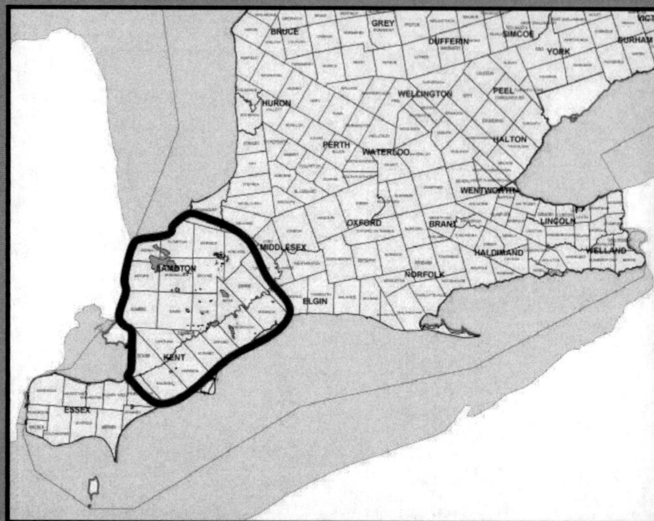

Devonian Pool Fairway (Lazorek, M., and Carter, T.R., 2008)

PLAY BY PLAY
EXPLORING THE MICHIGAN BASIN

9

Devonian Pool Location Map

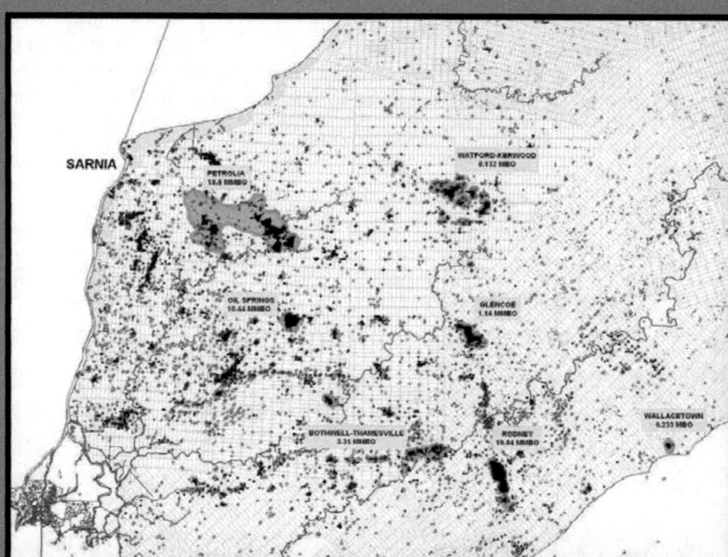

PLAY BY PLAY
EXPLORING THE MICHIGAN BASIN

10

Dundee Structure Map

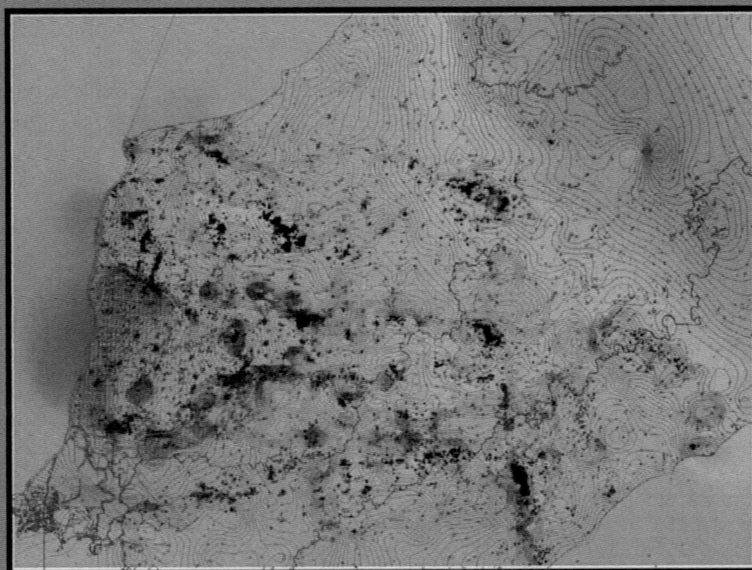

PLAY BY PLAY
EXPLORING THE MICHIGAN BASIN

11

Salt-Dissolution Generated Structural Traps

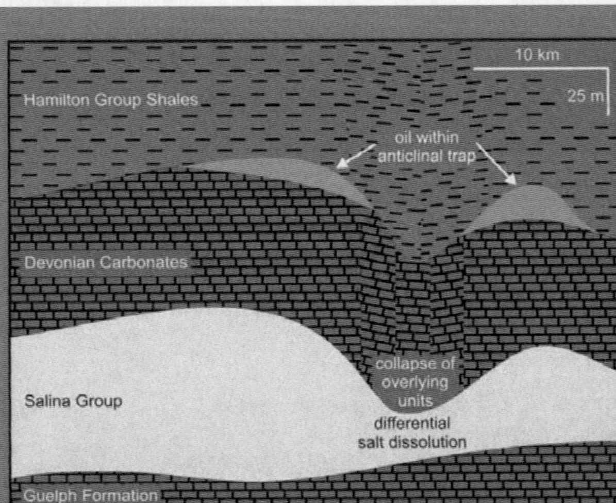

Conceptual model of Devonian structural traps formed by differential salt dissolution in the Salina Group (Lazorek, M., and Carter, T.R., 2008).

PLAY BY PLAY
EXPLORING THE MICHIGAN BASIN

12

Summary of Play Parameters

Key Reservoirs:

Dundee Formation :	Siliclastic-rich dolomite (Rodney) (from 100% sand to 100% dolomite)
Lucas Formation:	Sucrosic Dolomite (Oil Springs, Petrolia) Sandstone (Glencoe)
Other Production/Shows:	Dundee fractured limestones Hamilton fractured shale
Porosity:	8 to 30%
Permeability:	5 to 2,500 mD
Trapping Seal:	Hamilton shale, tight Dundee limestone

PLAY BY PLAY
EXPLORING THE MICHIGAN BASIN

13

Summary of Play Parameters

Trapping Mechanism:	Salt-Solution generated structural highs
Depth of Reservoirs:	350 to 450 feet
Thickness of Overburden:	50 to 300 feet
Exploration Methods:	Oil Seeps, tar beds, structural testing
Drilling and Casing:	Mainly cable tool and minor rotary Open-hole and cased-through and perforated
Completion:	Acidization and some frac'ing

PLAY BY PLAY
EXPLORING THE MICHIGAN BASIN

14

Rodney Field Example

General Field Data

Date of Discovery:	1949
Exploration Methods:	Structural Testing
Type of Trap:	Structural (Salt-generated)
Reservoir Lithology:	Siliclastic-rich dolomite
Producing Area:	1600 acres
Number of Producing Wells:	200 Producers, 100 Injectors
Drilling and Casing:	Open-hole and cased-through and perforated
Completion:	Acidization and some frac'ing

PLAY BY PLAY
EXPLORING THE MICHIGAN BASIN

15

Rodney Field Example

Reservoir Data

Producing Formation:	Dundee "Columbus" Zone
Depth to Reservoir:	360 to 430 feet
Deepest Zone Tested:	Cambrian
Pay Thickness:	15 to 24 feet (4 to 6 meters)
Pay Lithology	Siliciclastic-rich Dolomite
Pore Types:	Intercrystalline, vuggy, fracture and intergranular
Porosity:	10 to 30%
Permeability:	5 to 2,500 mD
Oil Character:	38 API (sour)
Water Saturation:	Variable 10 to 30%

PLAY BY PLAY
EXPLORING THE MICHIGAN BASIN

16

Rodney Field Example

Other Reservoir Data

Original Reservoir Pressure:	175 psig (1,206 kPa)
Drive Mechanism:	Solution Gas
Original OOIP:	18 million barrels
Production to Date:	10.845 million barrels
Initial Production/Well:	2 to 118 BOPD
Peak Daily Production:	1,200 BOPD
Current Daily Production:	60 BOPD
Secondary Recovery:	Water-flood on 5-spot, line and peripheral pattern
Cores:	10
Wireline Logs:	Basic Gr/N to Nil.

PLAY BY PLAY
EXPLORING THE MICHIGAN BASIN
17

Dundee Structure Map Rodney Pools

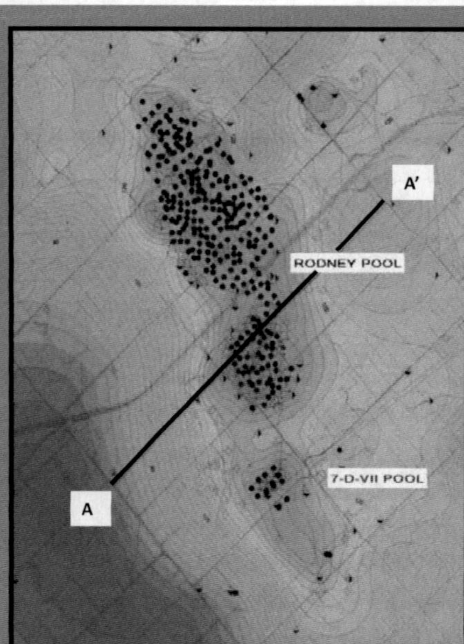

PLAY BY PLAY
EXPLORING THE MICHIGAN BASIN
18

Rodney Pool 2D Seismic Template Line

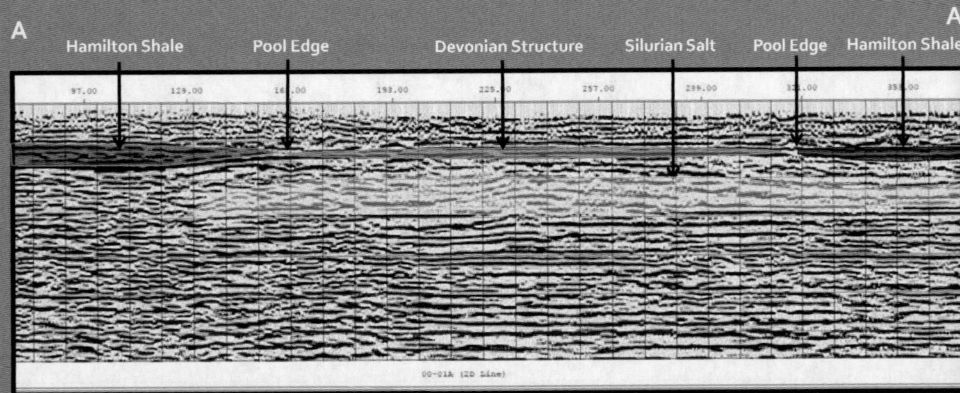

PLAY BY PLAY
EXPLORING THE MICHIGAN BASIN
19

Rodney Field Type Log

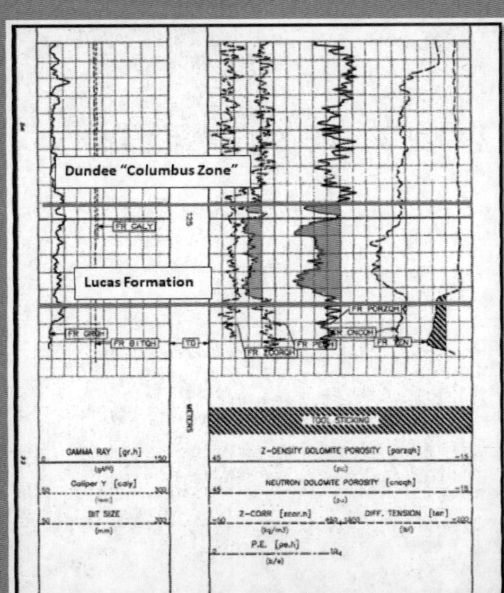

- 12 to 24% porosity
- Quartz-sand-rich intervals
- Dolomitized crinoidal grainstone to packsone
- 20 foot (6 m) pay section

PLAY BY PLAY
EXPLORING THE MICHIGAN BASIN
20

Use of Modern Technology

Exploration Tools

- Computer mapping of key geological tops and isopachs
- High-resolution gravity and magnetics
- Potential mapping of structures with 2D (3D) seismic

Waterflooding

- Successful water-flood results in the Rodney field
- 5-spot flood pattern very effective
- To date recovered approximately 60% of OOIP

PLAY BY PLAY
EXPLORING THE MICHIGAN BASIN

21

Use of Modern Technology

Horizontal Drilling

- To date marginal success based on 3 wells
- Challenge to make the turn within 100 to 150 feet bedrock
- Relatively thin pay zone and heterogeneous reservoir
- Vertical permeability variable
- Due to shallow depth, lack of significant reservoir drive
- Limited stimulation options due to proximity of overlying fresh water aquifer and underlying Lucas water zones

PLAY BY PLAY
EXPLORING THE MICHIGAN BASIN

22

Future Potential

- Remaining potential for smaller pools in 1 to 5 million barrels
- Areas within the Devonian fairway with relatively sparse drilling
- Potential to image structures with high-resolution gravity due to density contrasts of preserved sections of Salina salts.
- Use of magnetics data (OGS dataset) to map potential faults
- Use of old-fashioned conventional geological mapping
- Use of 2D seismic to confirm structure and drill locations
- Attractive economics with relatively low cost to drill a vertical test well $75,000 to $80,000

Future Potential Exploration Areas

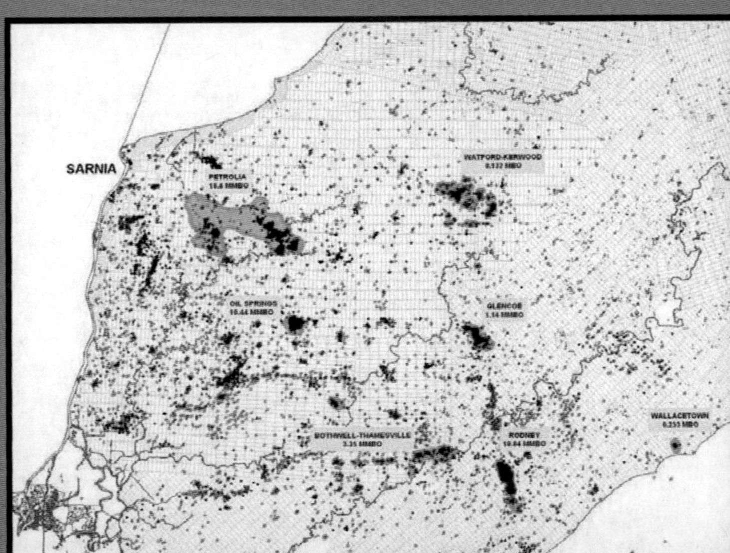

DEVONIAN OIL AND GAS PLAYS OF THE MICHIGAN BASIN

William B. Harrison, III

Michigan Geological Repository for Research and Education (MGRRE)

Western Michigan University-Michigan Geological Survey

PLAY BY PLAY

Exploring the Michigan Basin

1

Devonian Hydrocarbon Production in Michigan

- Commercial development began in the 1920's with Traverse and Dundee Oil
- Detroit River Group oil and gas production began in the mid-1930's
- Antrim Shale and Amherstburg production took off in the 1980's
- Antrim Shale fields are defined as "continuous" and named by their County - over 12,000 wells have been drilled in the Antrim Shale play – all are gas producers. Currently around 9000 wells are producing.
- Traverse Lime reservoirs are found in 265 classified fields – 96 are Gas fields, 169 are oil fields
- Dundee producing fields include those with production in the Dundee, Rogers City and Reed City units. There are 171 classified fields – 42 are Gas, 128 are oil and 1 is Gas Storage
- Detroit River Group production comes from the "Detroit River *sour zone*" in the Horner Member of the Lucas Fm. and the Richfield Member of the Lucas Fm. There are 114 classified fields in this interval.
- Amherstburg production comes from 5 known fields

PLAY BY PLAY
EXPLORING THE MICHIGAN BASIN

2

Hydrocarbon Producing Devonian Rocks in Michigan

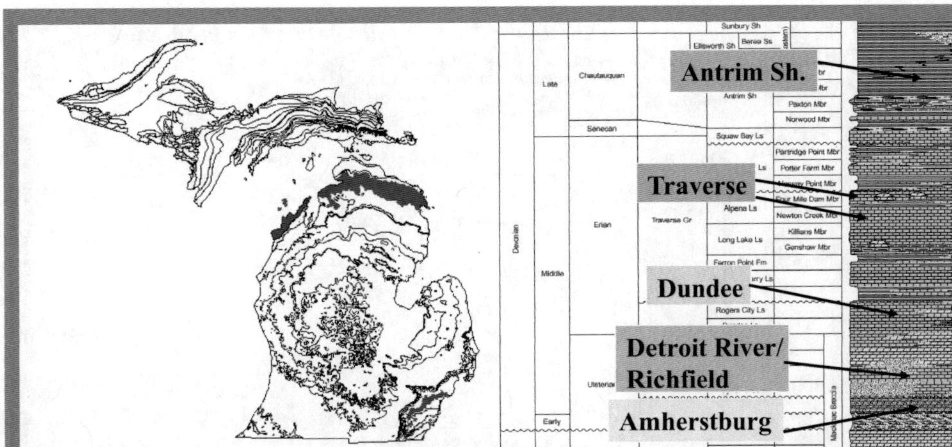

Devonian reservoirs include Antrim Shale,Traverse Limestone, Dundee Formation, Detroit River Group and Amherstburg Fm. Outcrop and Subcrop of these rocks shown in blue.

Modified from Michigan Geological Survey Presentation–"The Rock Cycle"

PLAY BY PLAY
EXPLORING THE MICHIGAN BASIN
3

MAP OF OIL AND GAS WELLS FOR MICHIGAN'S DEVONIAN RESERVOIR PLAYS

From State of Michigan Geological Survey, 2003

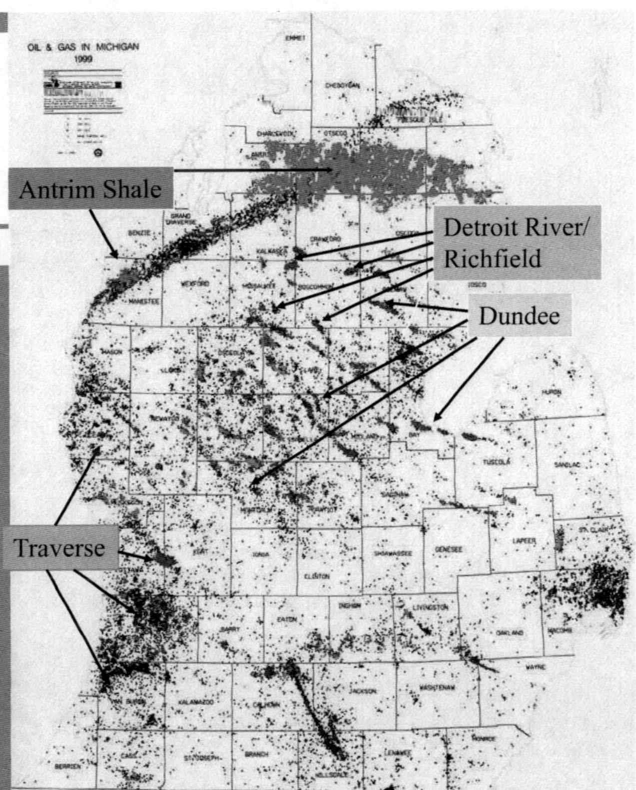

PLAY BY PLAY
EXPLORING THE MICHIGAN BASIN
8

Antrim Shale Production in Michigan

- Antrim Shale production began in the 1940's, but large-scale, commercial began in the 1980's
- Production peaked in 1998
- Cumulative production through 2016 is 3,560,645,741 MCF of natural gas.
- Two wells in 2015 produced 1115 bbls oil from central basin

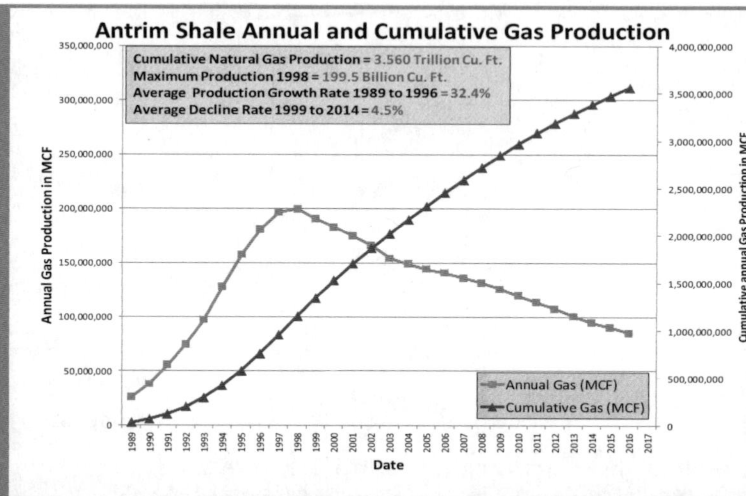

PLAY BY PLAY
EXPLORING THE MICHIGAN BASIN
5

ANTRIM SHALE – LACHINE MEMBER

Welch-St. Chester 18, Otsego County

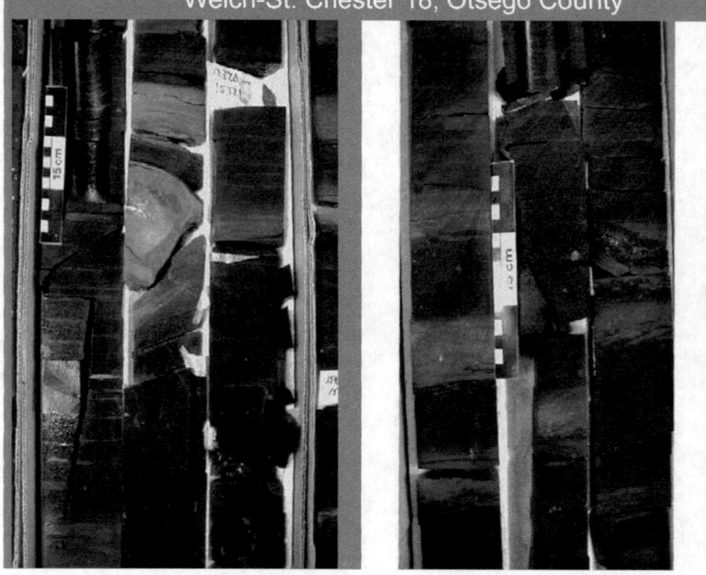

- The Antrim Shale is an organic-rich (up to 25% TOC
- There is moderate clay content and high silica content
- Extensive fracturing may be present from a combination of glacial unloading and rebound and deeper tectonic structures
- Fresh water recharge into the fractures facilitates methanogenic bacteria

PLAY BY PLAY
EXPLORING THE MICHIGAN BASIN
6

Michigan Basin during Middle Devonian Time

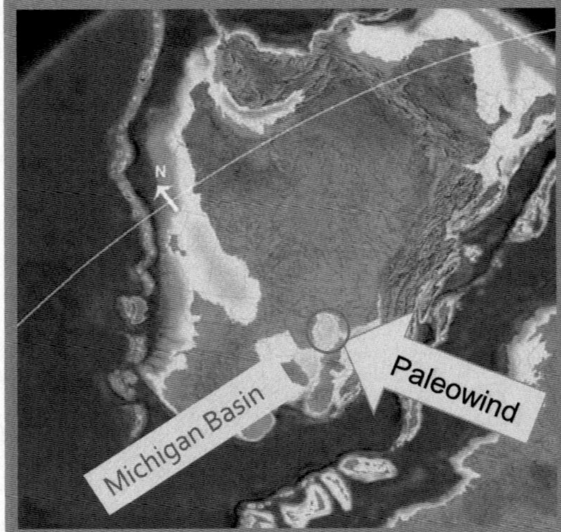

- Located about 25 degrees south of the Equator in the tropical trade winds belt
- Shallow marine and peritidal depositional environments
- Predominately carbonate systems with low clastic input most of the time

TRAVERSE LIME STRATIGRAPHY

- Traverse Limestone is defined from outcrops and in Quarries in northern Lower Michigan
- "Traverse Lime" is general subsurface name for entire carbonate strata between Base of the Antrim Shale and the top of the Dundee/Rogers City
- Many formations and members have been named in outcrop, but are not distinguished in subsurface
- Production is near the top of the package

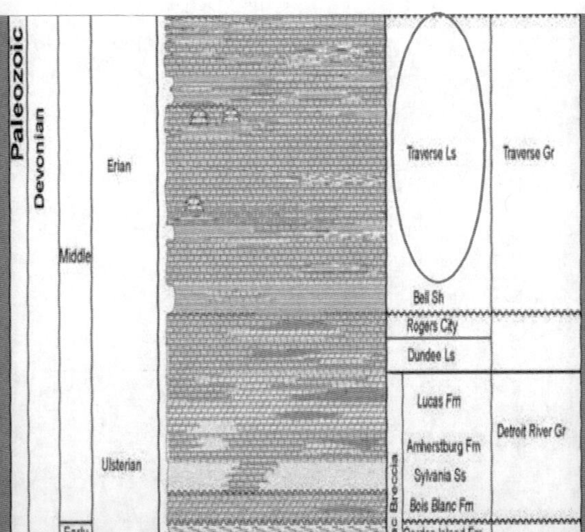

(modified from Catacosinos et al., 1990)

Traverse Lime Production in Michigan

- Production from individual Devonian carbonate reservoirs cannot be precisely documented because historically production totals have been comingled from the Traverse, Dundee and Detroit River Groups in some fields. Additionally, no publicly available production data exists from 1986 to 1997. That 10 year period can only be estimated from decline curves.
- Traverse Lime has produced since the mid-1920's, but many field discoveries were made in the 1930's and 1940's. Cumulative production from the Traverse Lime play exceeds 110 million barrels of oil

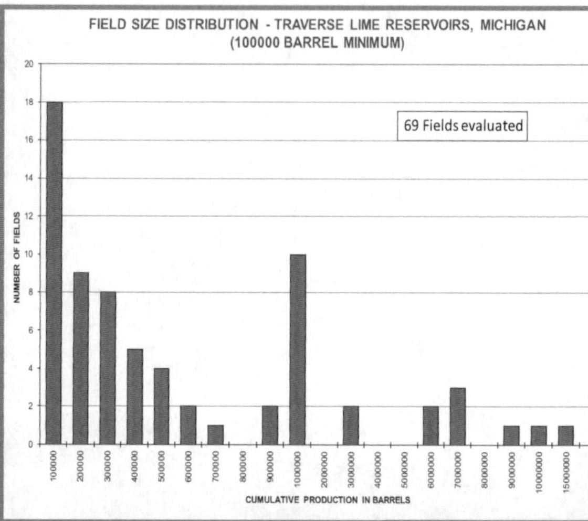

PLAY BY PLAY
EXPLORING THE MICHIGAN BASIN
9

TRAVERSE LIMESTONE RESERVOIRS

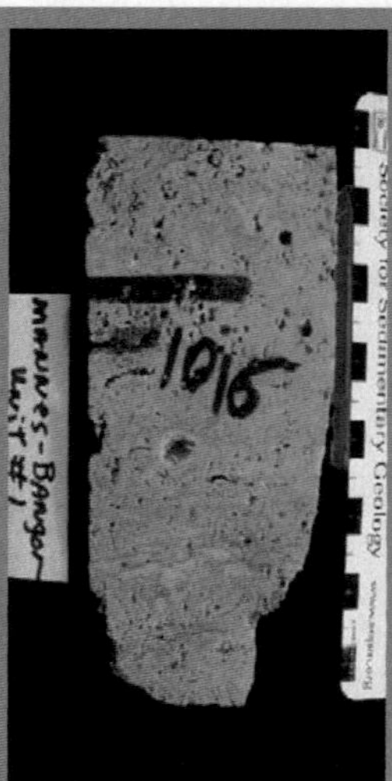

PLAY BY PLAY
EXPLORING THE MICHIGAN BASIN
10

Scale of Michigan compared to Modern Bahama Platform (Same Scale on Both)

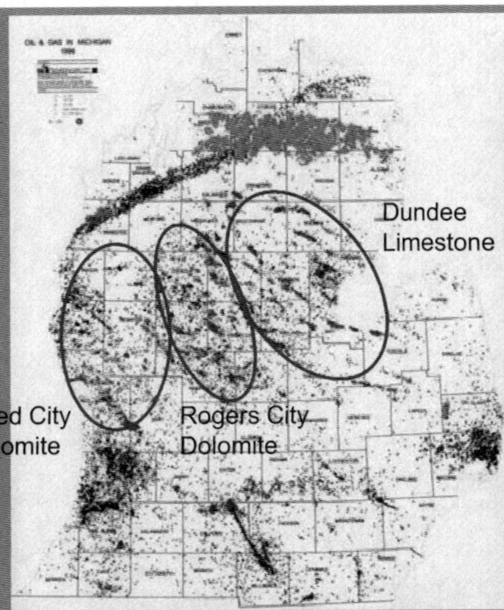

Michigan Oil and Gas Well Location Map

PLAY BY PLAY
EXPLORING THE MICHIGAN BASIN

DUNDEE AND ROGERS CITY STRATIGRAPHY

- Rogers City and Dundee Formations are defined from exposures in Quarries in Presque Isle Co.
- "Dundee" is general subsurface name for entire carbonate strata between Lucas Fm. and Bell Shale
- Both can be distinguished in subsurface

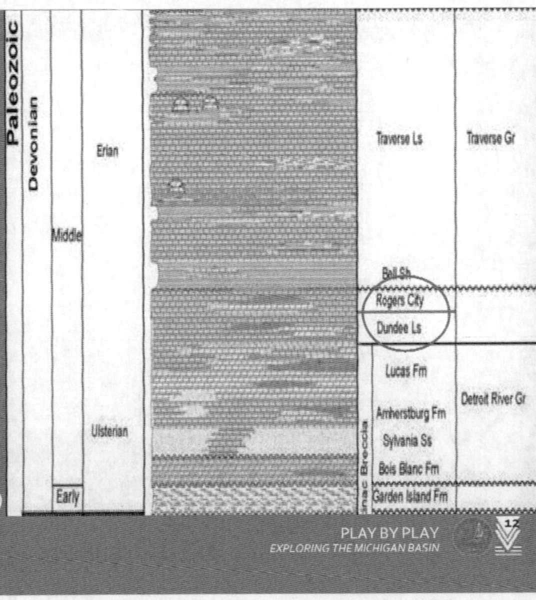

(modified from Catacosinos et al., 1990)

PLAY BY PLAY
EXPLORING THE MICHIGAN BASIN

Dundee Production in Michigan

- Dundee production comes from three separate horizons in the unit. The Rogers City Fm. directly underlies the Bell Shale and is productive only when altered to Dolomite. The Reed City Member of the Dundee is a restricted dolomite unit below an Anhydrite member in the western part of the basin.
- Cumulative oil production more than 360 million Bbls from 1928-1986 and 1997-2015 *(1987 to 1996 production from the Dundee and Reed City is not reported, estimated 15-20 additional BBLS)*
- First discovery in 1928 (Mt Pleasant Field)
- Most fields discovered in 1930's and 1940's
- Reservoirs in both limestone and dolomite

Mt Pleasant Oil Field - ca. 1930's
Courtesy of archives at Clarke Historical Library, Central Michigan University

PLAY BY PLAY
EXPLORING THE MICHIGAN BASIN
13

NUMBER OF DUNDEE FIELDS DISCOVERED PER YEAR

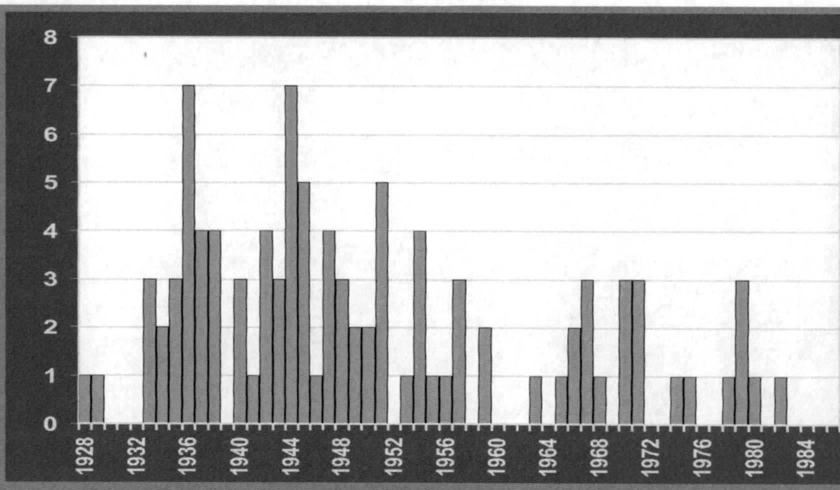

PLAY BY PLAY
EXPLORING THE MICHIGAN BASIN
14

DUNDEE FIELD SIZES BY DISCOVERY YEAR

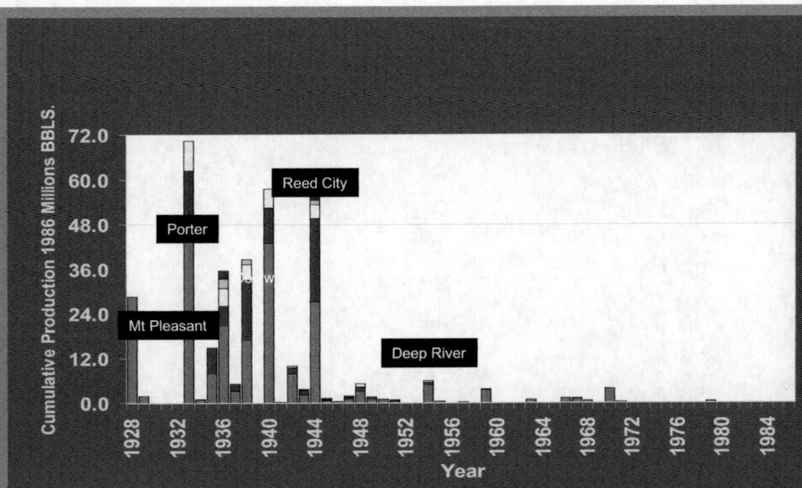

DUNDEE RESERVOIR LITHOLOGIC TYPES

- Limestone - depositional fabrics
  - Grainstone shoal sand bodies
  - Fenestral peritidal packstones and wackestones
  - Stromatoporoid/coral boundstone patch reefs
- Dolomite - burial diagenetic fabrics
  - Fractured, vuggy, hydrothermal dolomite
  - Matrix, intercrystalline and vuggy dolomite
  - Sucrosic, laminated, dolomite with anhydrite

DUNDEE RESERVOIR DEPOSITIONAL FABRICS

Reed City	Dundee	Rogers City
MichCon-LoReed #LR 83-2 Osceola Co.	Wiser-Sturm #4-0 Gladwin Co.	Thelma Rousseau #1-12 Mecosta Co.

Laminated and Fenestral Wackestone and Mudstone	Reef and Reef Skeletal Debris	Skeletal Packstone

PLAY BY PLAY
EXPLORING THE MICHIGAN BASIN
17

LITHOLOGIC PROPERTIES IN DUNDEE LIMESTONE RESERVOIRS

- Limestone reservoirs have significant original depositional fabrics preserved
- Porosity is primarily intergranular or intragranular
- Burial diagenesis has added some cements and produced chemical compaction (stylolites)
- Fracturing may or may not play a role

PLAY BY PLAY
EXPLORING THE MICHIGAN BASIN
18

LITHOLOGIC PROPERTIES IN DUNDEE DOLOMITE RESERVOIRS

- Dolomite reservoirs have pervasive replacement of original sediment
- Some original depositional fabric is preserved and provides a template for additional diagenesis
- Solution enhanced pores and vugs are common, often with HTD cements
- Intercrystalline porosity important
- Fractures usually occur and may be abundant

PLAY BY PLAY
EXPLORING THE MICHIGAN BASIN 19

DOLOMITE-LINED SOLUTION ENLARGED VUGGY POROSITY

REIMAN #1, Osceola County, MI

- Core from Reiman #1 well in Cat Creek Field, Osceola Co., MI
- White "Saddle" Dolomite lining pores
- Matrix is fine to medium crystalline planar dolomite
- Intercrystalline porosity up to 5%

PLAY BY PLAY
EXPLORING THE MICHIGAN BASIN 20

DUNDEE FIELDS MAP WITH CORED WELLS

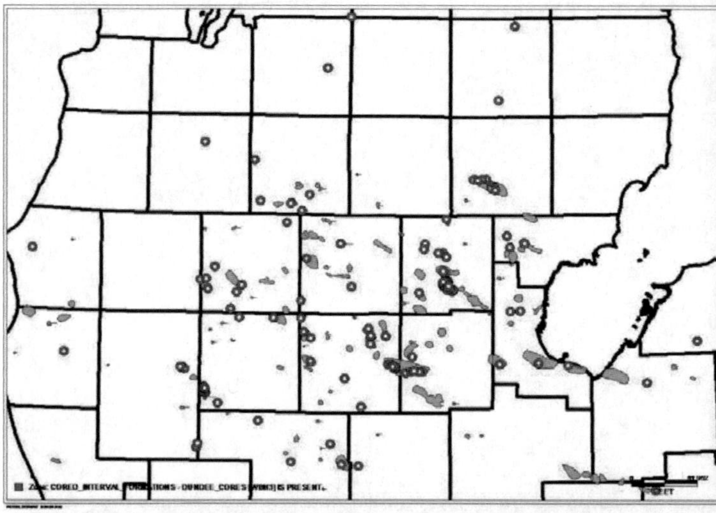

- 134 core wells in Dundee, Rogers City or Reed City intervals
- Cores represent all three zones of Dundee production
- Many are in Producing fields, but some are dry holes

PLAY BY PLAY
EXPLORING THE MICHIGAN BASIN

DUNDEE DRIVE MECHANISMS-PRIMARY AND SECONDARY RECOVERY FACTORS

- Limestone reservoirs have porous facies generally only on structures, relatively tight off structure – main drive is solution gas/gas expansion – is pressure in reservoir depletes, primary recovery will be low - good secondary potential
- Dolomite reservoirs have extensive porosity off structure because of regional dolomitization – main drive is bottom or lateral water drive – reservoir pressure does not deplete very much – limited secondary recovery potential

PLAY BY PLAY
EXPLORING THE MICHIGAN BASIN

ANNUAL OIL AND WATER PRODUCTION WEST BRANCH FIELD - LIMESTONE - GRAINSTONE

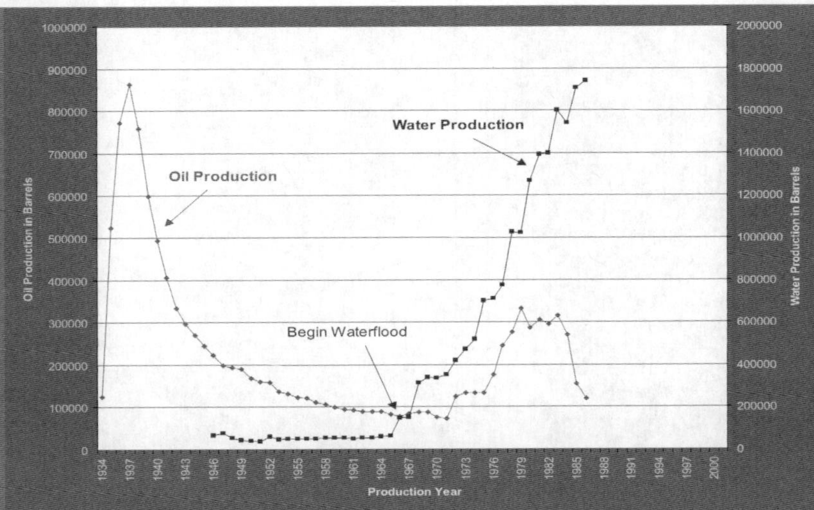

PLAY BY PLAY
EXPLORING THE MICHIGAN BASIN
23

ANNUAL OIL AND WATER PRODUCTION DEEP RIVER FIELD – FRACTURED DOLOMITE

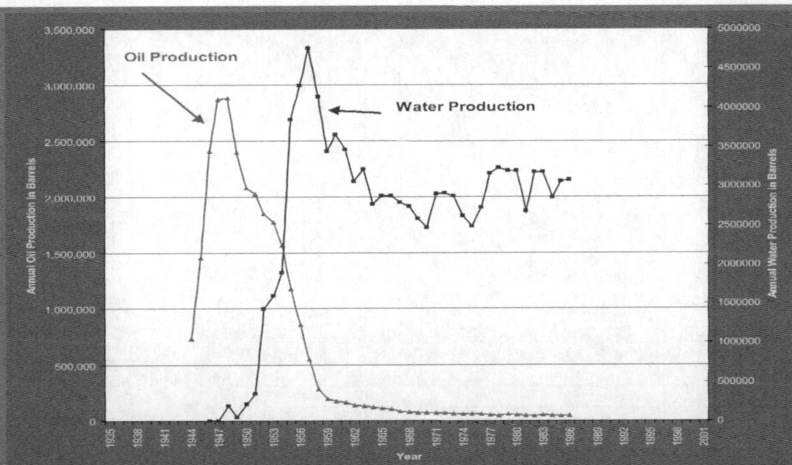

PLAY BY PLAY
EXPLORING THE MICHIGAN BASIN
24

LUCAS AND AMHERSTBURG STRATIGRAPHY

- Rogers City and Dundee Formations are defined from exposures in Quarries in Presque Isle Co.
- "Dundee" is general subsurface name for entire carbonate strata between Lucas Fm. and Bell Shale
- Both can be distinguished in subsurface

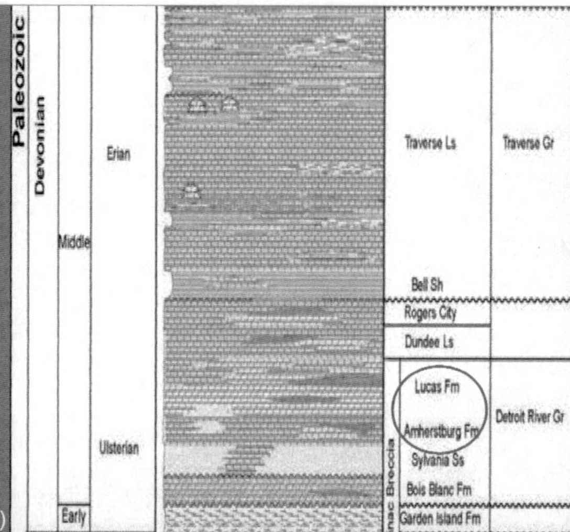

(modified from Catacosinos et al., 1990)

PLAY BY PLAY
EXPLORING THE MICHIGAN BASIN
25

Detroit River Group Production

- Detroit Rive Group Produces from two zones in restricted, sucrosic Dolomite, "sour Zone" in the Horner member of the Lucas Fm. and the Richfield Member of the Lucas Fm.
- 94 fields are classified as Oil fields – some fields produce from both zones, some from only one zone
- Cumulative production from the two zones is in excess of 100 million barrels of oil and 89 BCF of gas
- Waterflooding operations have been initiated in 20 of these fields, 10 in the Richfield and 10 in the Detroit River "sour zone" or in both units.
- Top 10 fields have produced 86% of the play's oil.

Field Name	County	Total wells	Actve Wells	Field Acres	Cum Oil BBLS	Cum Gas MCF
Beaver Creek	Crawford	138	84	4960	24,080,332	21,102,461
East Norwich	Missaukee	257	188	5760	17,000,000	17,000,000
Rose City	Ogemaw	134	70	5360	11,500,000	12,000,000
St Helen	Roscommon	153	101	4600	9,000,000	15,000,000
Hamilton	Clare	47	26	1870	6,000,000	3,000,000
Enterprise	Missaukee	76	55	1760	5,500,000	2,500,000
Reed City	Osceola	45	21	1800	5,000,000	4,000,000
Rich	Lapeer	58	31	2160	3,500,000	5,500,000
Headquarters	Roscommon	70	41	2320	3,000,000	2,500,000
Kawkawlin	Bay	21	20	780	2,500,000	1,500,000

PLAY BY PLAY
EXPLORING THE MICHIGAN BASIN
26

RICHFIELD MEMBER OF THE LUCAS FM. RESERVOIR AND INTERNAL ANHYDRITE SEALS

Dart-Porter Hogan #1-17, Roscommon County

PLAY BY PLAY
EXPLORING THE MICHIGAN BASIN
27

Amherstburg Fm. – Source Rock and Reservoir

Core Energy-St Charlton #4-30, Otsego Co.

Hunt-McGuire #1-22, Oscoda Co.

PLAY BY PLAY
EXPLORING THE MICHIGAN BASIN
28

Summary and Conclusions

- Devonian Reservoirs in Michigan represent about 45 % of the oil production and 50% of the natural gas production all time from the State.
- Total oil production exceeds 585 million barrels, Gas production exceeds 3.8 TCF
- Reservoirs are mostly found in Carbonate rocks, with dolomitized reservoirs generally be higher quality, but some Limestone zones maintain porosity and permeability due to specific original depositional facies
- The Antrim Shale is the only non-Carbonate producing zone in the Devonian of Michigan
- Secondary recovery has been attempted in the oil reservoirs of several plays with good results in some units and poor performance in some others

TRENTON- BLACK RIVER HTD RESERVOIRS OF THE MICHIGAN BASIN

Ian Colquhoun – Ontario

Bill Vansickel - Michigan

PLAY BY PLAY

Exploring the Michigan Basin

1

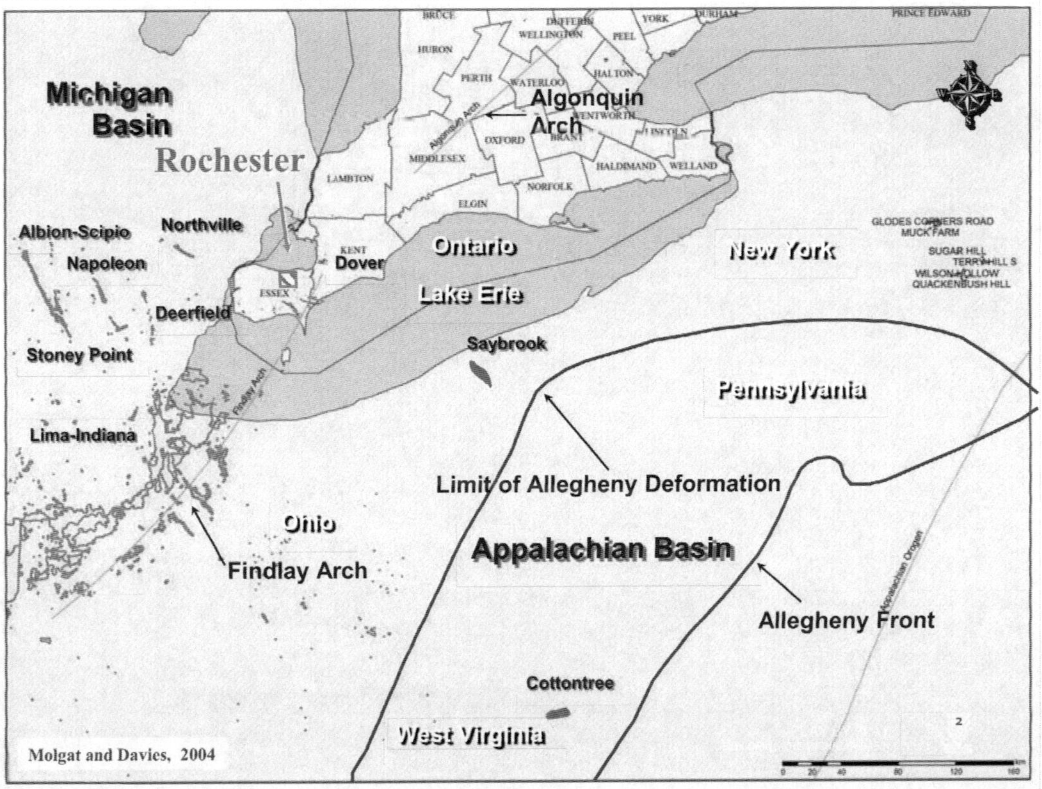

2

ONTARIO ORDOVICIAN POOLS AND RECOVERABLE RESERVES

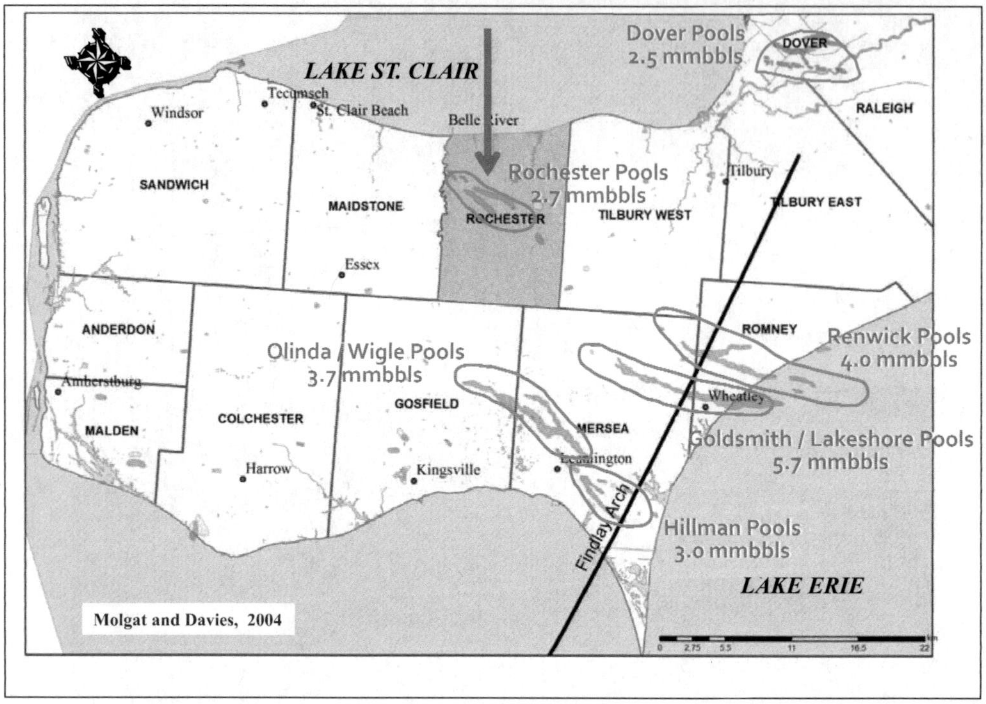

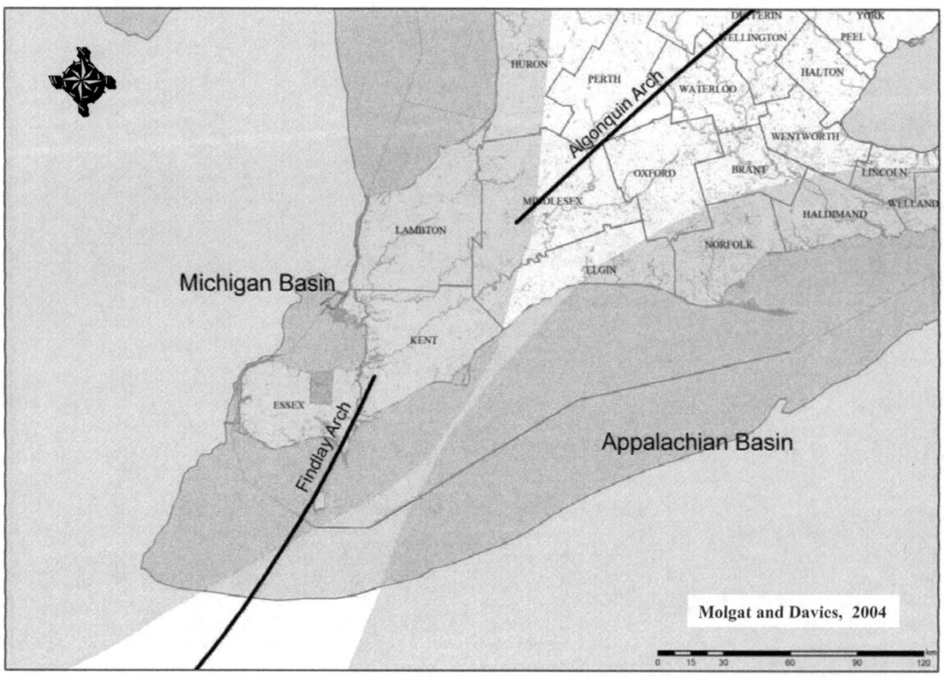

General Stratigraphic Section for SW Ontario

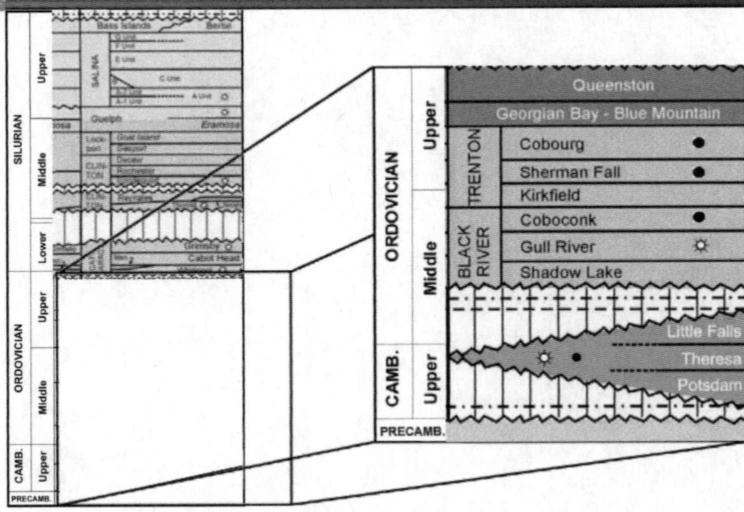

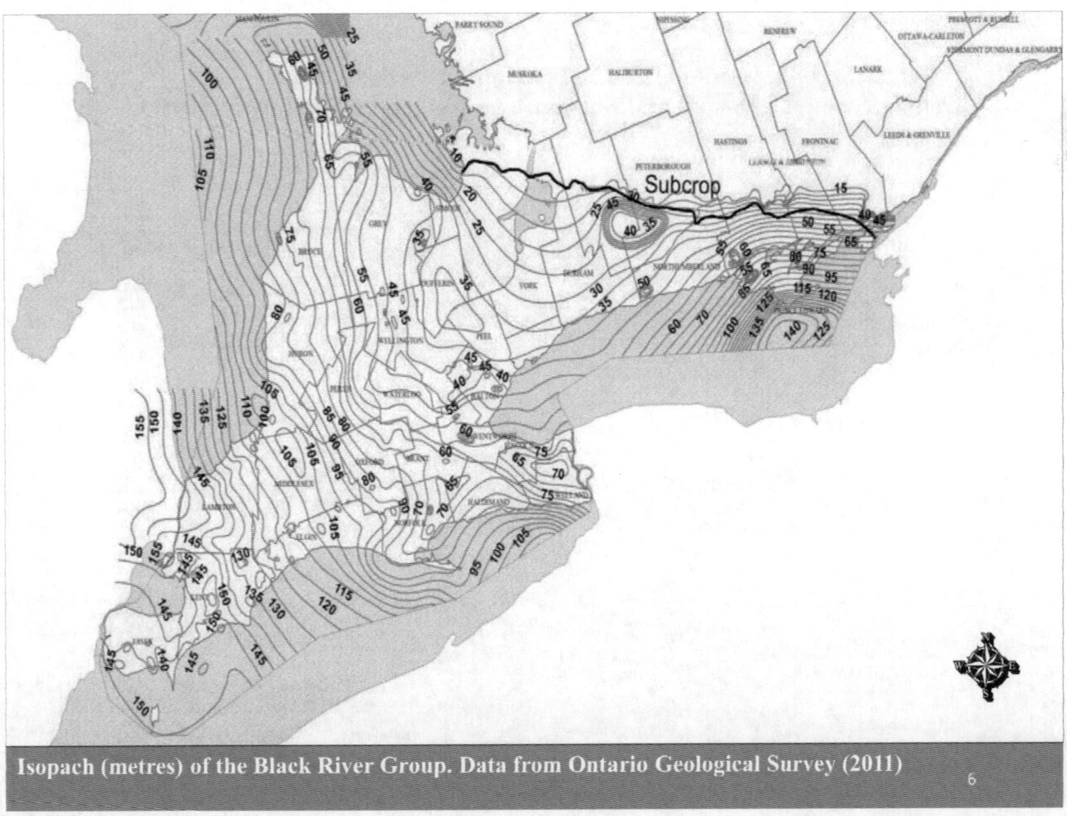

Isopach (metres) of the Black River Group. Data from Ontario Geological Survey (2011)

6

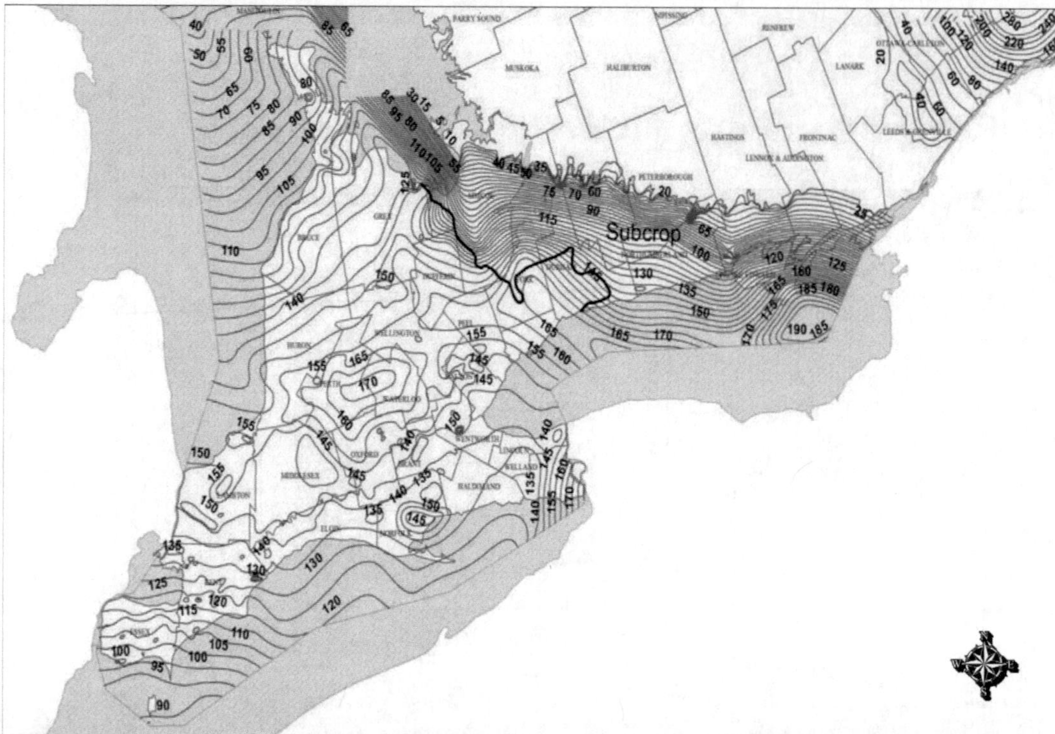

Isopach (metres) of the Trenton Group. Data from Ontario Geological Survey (2011).

7

Generalized Depositional Facies Distribution of the Trenton Group Carbonates within Southwestern Ontario

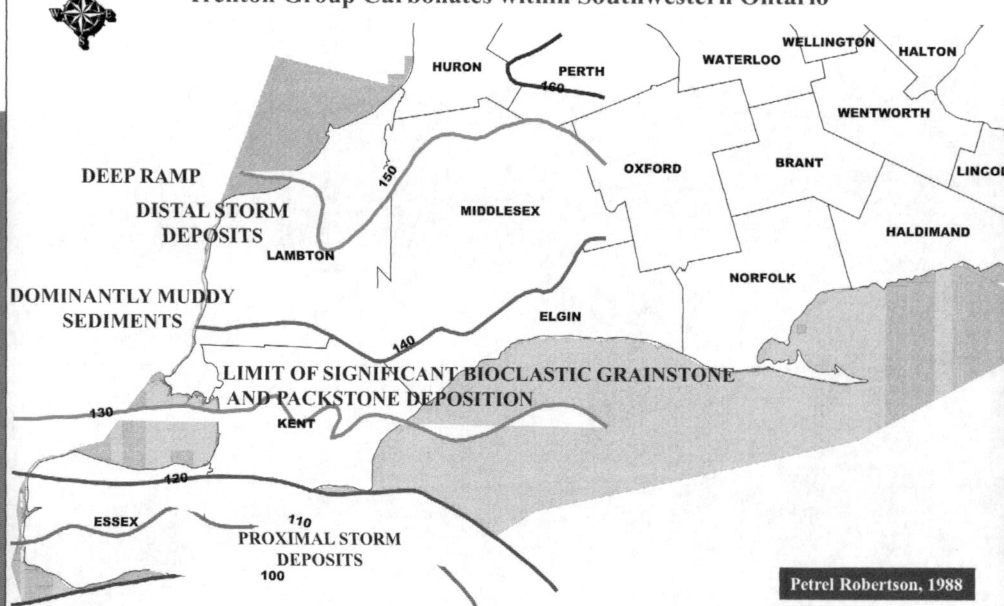

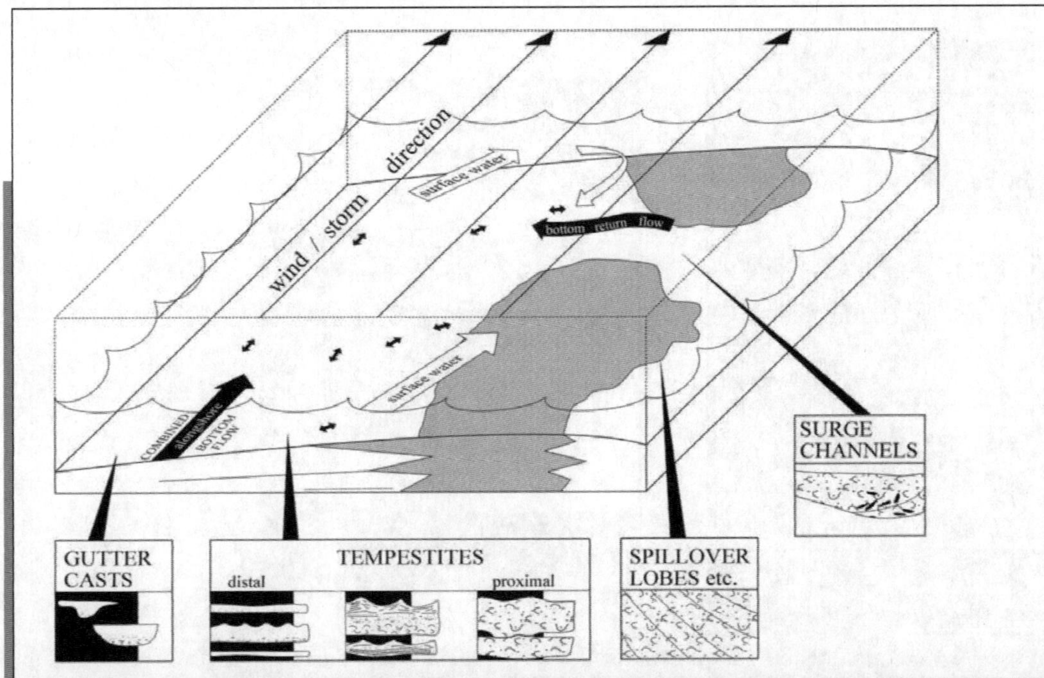

Crinoidal Ramp Facies Model showing the ramp dynamics and resulting facies distribution (Aigner 1985)

9

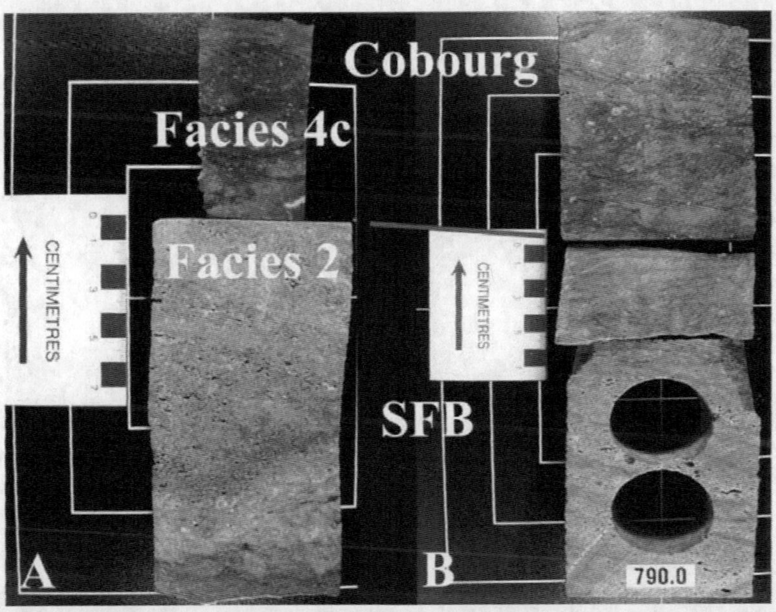

Core Photos 1A, B: Note the sharp boundary between facies 4c within the Cobourg and facies 2 at the top of the Sherman Fall Bioclastic. A) 2238 - 2238.5 feet in British American #19, Malden 4-67-VI, and B) 789.65 - 790.0 mKB in Consumers' et al Mersea 3-12-I.

10

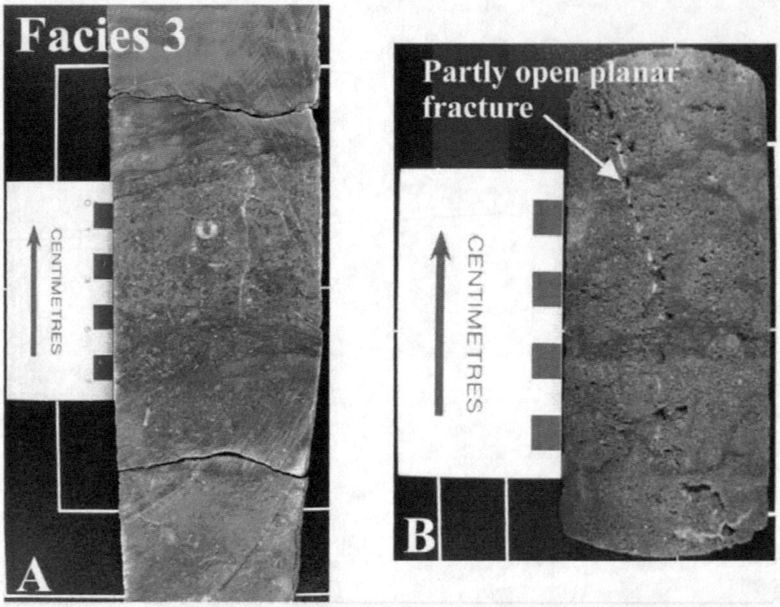

Core Photos 2 A, B: Facies 3 observed in the Coboconk Formation from A) 1012.06 to 1012.30 mKB and from B) 1017.60 to 1017.70 mKB. Note the partly open fracture (core photo B) that is lined with dolomite rhombs and partly filled with clear to white calcite cement.

11

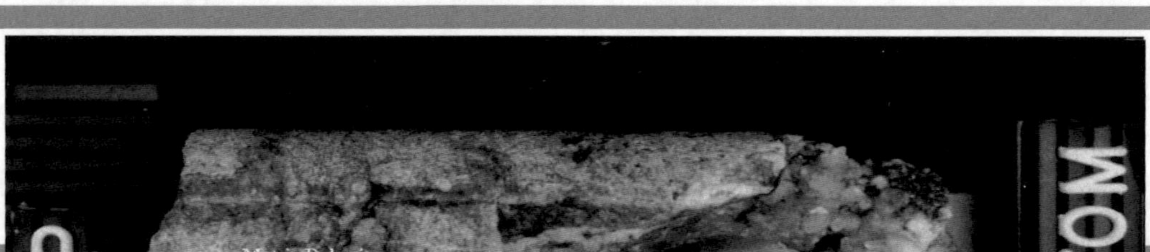

Dolomitization from relatively hot, Mg-rich fluids migrated through the regional porous Cambrian sandstone and upwards into Trenton / Black River strata, along a complex network of faults and fractures. The migrating fluids had the ability to dissolve and dolomitize the limestones and to precipitate porosity-occluding saddle dolomite, calcite and sulphides. The most pervasive dolomitization is developed in the Sherman Fall Bioclastic member because of its higher primary porosity.

12

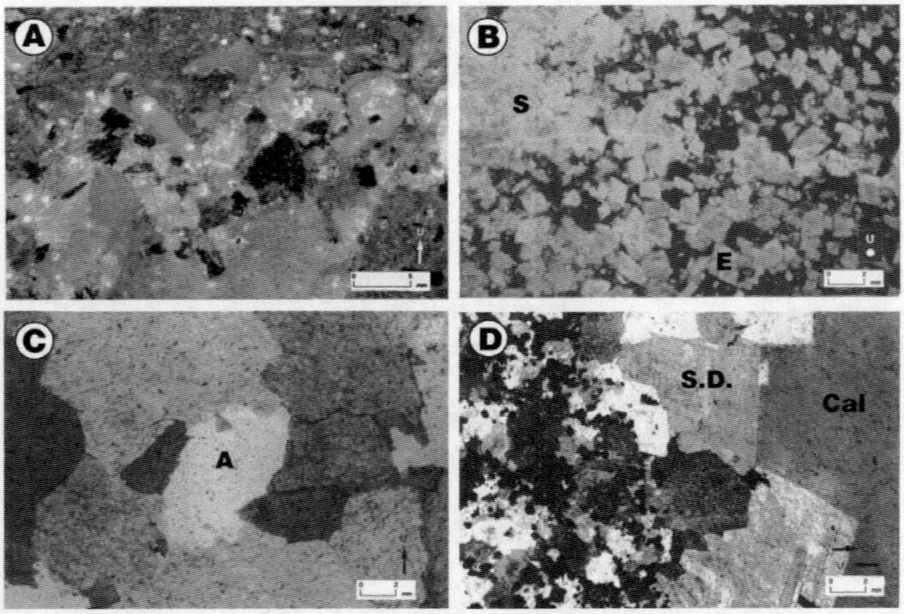

A) Photomicrograph of stained thin section under plane light. Dolomite rhombs occur interstitial to bioclastic material within the matrix. B) Photomicrograph of stained thin section under polarized light of planar-E dolomite. Note the gradation from porous planar-E dolomite (E) to finer-grained planar-S dolomite (S). C) Photomicrograph of stained thin section under plane light of nonplanar-A dolomite (A). Coarse-grained nonplanar dolomite crystals show rare development of intracrystalline porosity. D) Photomicrograph of stained section under plane light of coarse-grained, euhedral saddle dolomite (S.D.) cement (white; extinguishing grey) lining a large vug filled with equant blocky calcite (Cal) cement (red).

13

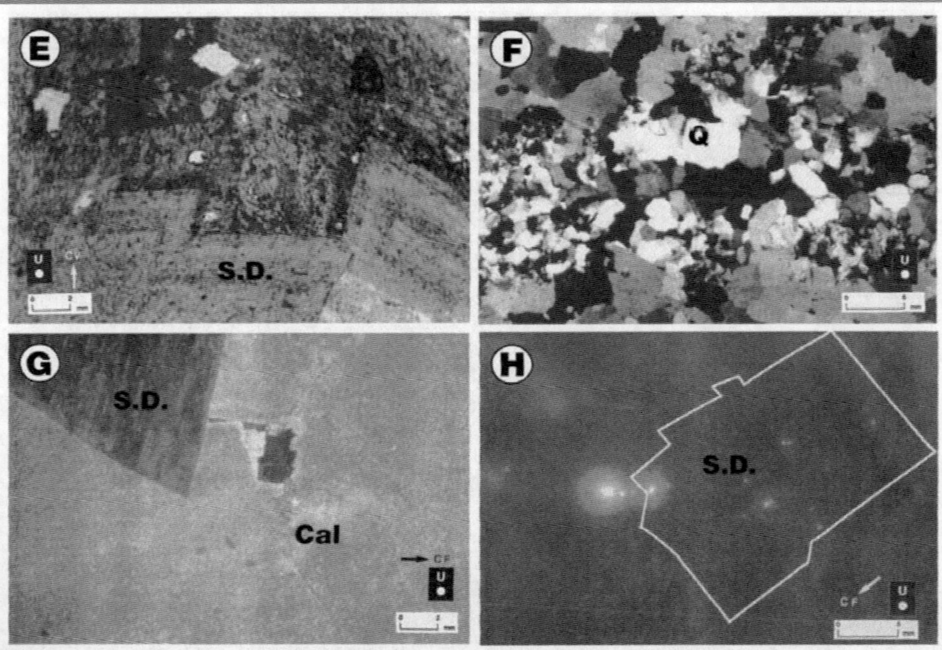

E) Photomicrograph of stained thin section under plane light. This example illustrates the replacement of saddle dolomite crystals to calcite with the complete infilling by calcite cement within the centre of a large vug. F) Photomicrograph of stained thin section under plane light. The coarse quartz grains increase in size toward the centre of the vug. G) Cathodoluminescence photomicrograph of a large saddle dolomite (S.D.) crystal within equant, blocky calcite (Cal) cement. H) Photomicrograph under ultraviolet fluorescence showing the boundary of earliest saddle dolomite (S.D.) cement within a fracture and fluorescence of hydrocarbons (yellow) in fluid inclusions. (U = up direction in core, CV = centre of vug, CF = centre of fracture).

14

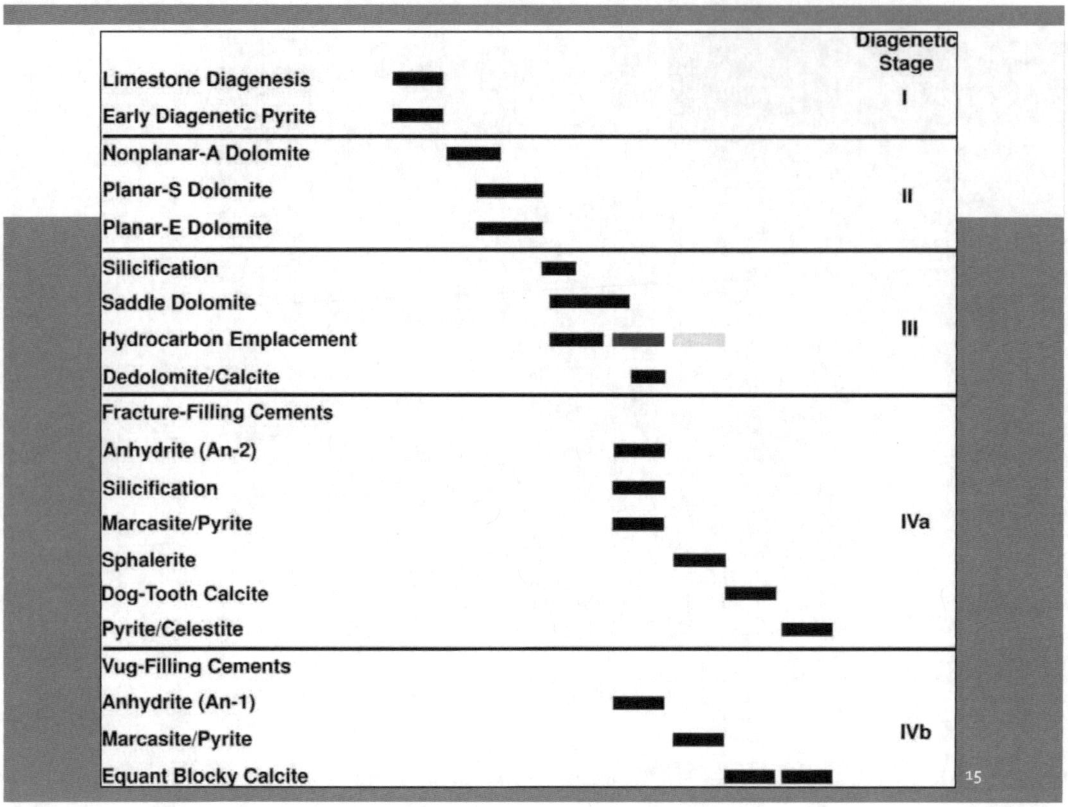
Diagenetic Stage
Limestone Diagenesis
Early Diagenetic Pyrite
I
Nonplanar-A Dolomite
Planar-S Dolomite
Planar-E Dolomite
II
Silicification
Saddle Dolomite
Hydrocarbon Emplacement
Dedolomite/Calcite
III
Fracture-Filling Cements
Anhydrite (An-2)
Silicification
Marcasite/Pyrite
Sphalerite
Dog-Tooth Calcite
Pyrite/Celestite
IVa
Vug-Filling Cements
Anhydrite (An-1)
Marcasite/Pyrite
Equant Blocky Calcite
IVb
15

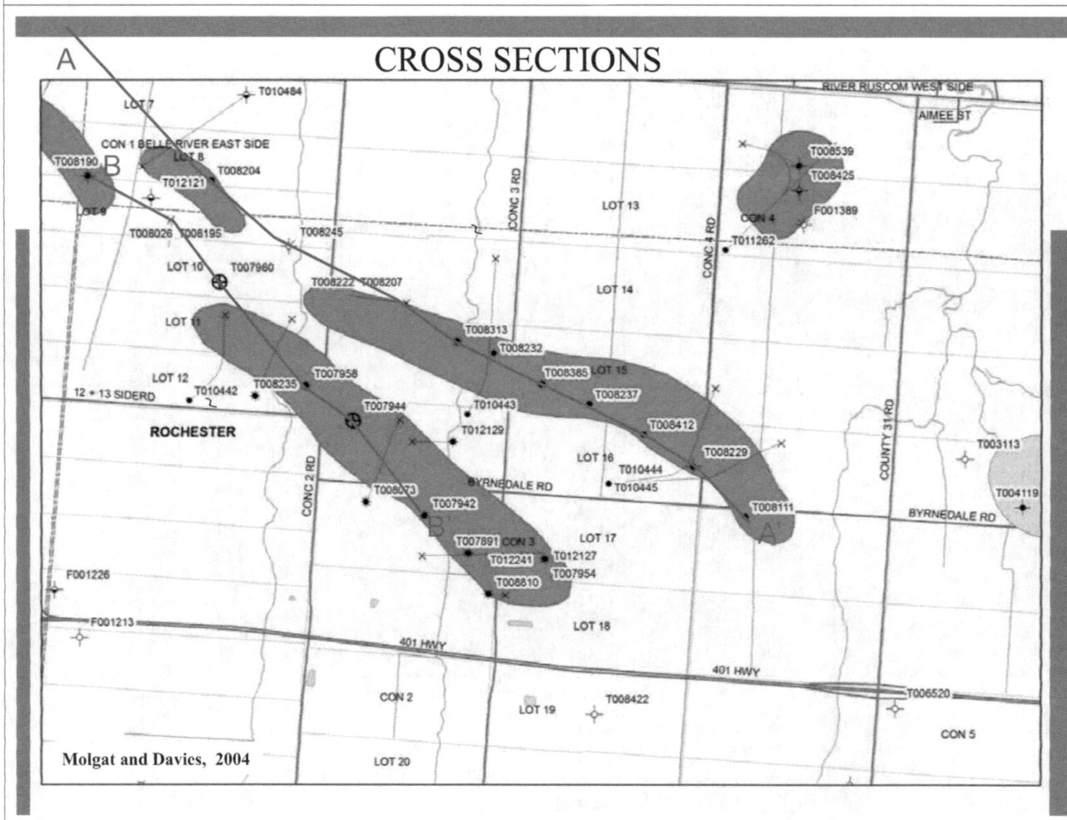
CROSS SECTIONS
A
B
B'
A'
ROCHESTER
CON 1 BELLE RIVER EAST SIDE
RIVER RUSCOM WEST SIDE
AIMEE ST
12 + 13 SIDERD
BYRNEDALE RD
CONC 2 RD
CONC 3 RD
CONC 4 RD
COUNTY 31 RD
401 HWY
CON 2
CON 3
CON 4
CON 5
LOT 7
LOT 8
LOT 10
LOT 11
LOT 12
LOT 13
LOT 14
LOT 15
LOT 16
LOT 17
LOT 18
LOT 19
LOT 20
Molgat and Davies, 2004

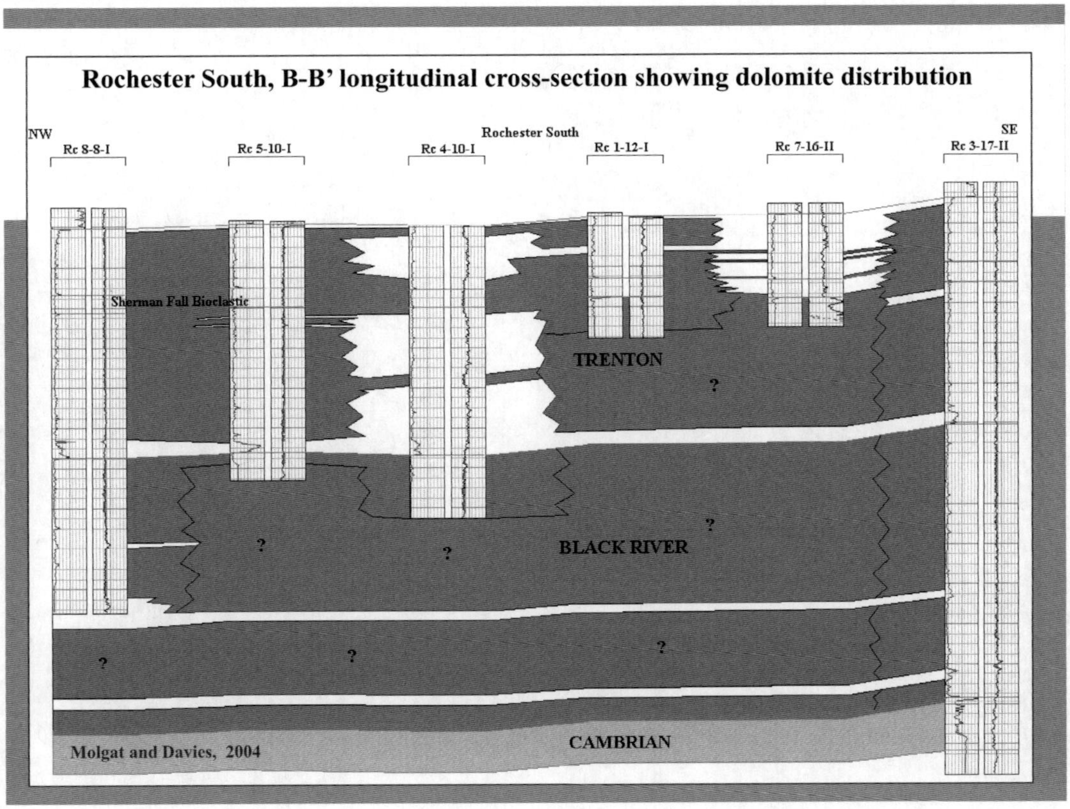
Rochester South, B-B' longitudinal cross-section showing dolomite distribution
NW
Rochester South
SE
Rc 8-8-I
Rc 5-10-I
Rc 4-10-I
Rc 1-12-I
Rc 7-16-II
Rc 3-17-II
Sherman Fall Bioclastic
TRENTON
BLACK RIVER
CAMBRIAN
Molgat and Davies, 2004

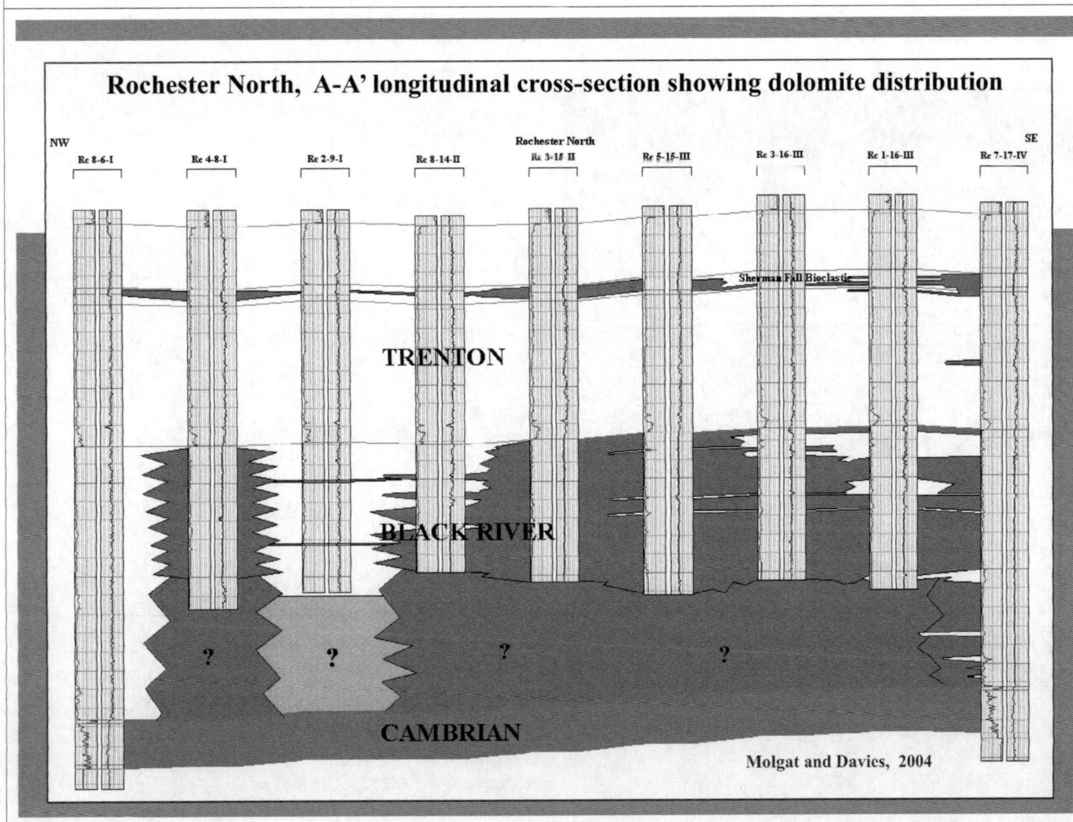
Rochester North, A-A' longitudinal cross-section showing dolomite distribution
NW
Rochester North
SE
Rc 8-6-I
Rc 4-8-I
Rc 2-9-I
Rc 8-14-II
Rc 3-15 II
Rc 5-15-III
Rc 3-16-III
Rc 1-16-III
Rc 7-17-IV
Sherman Fall Bioclastic
TRENTON
BLACK RIVER
CAMBRIAN
Molgat and Davies, 2004

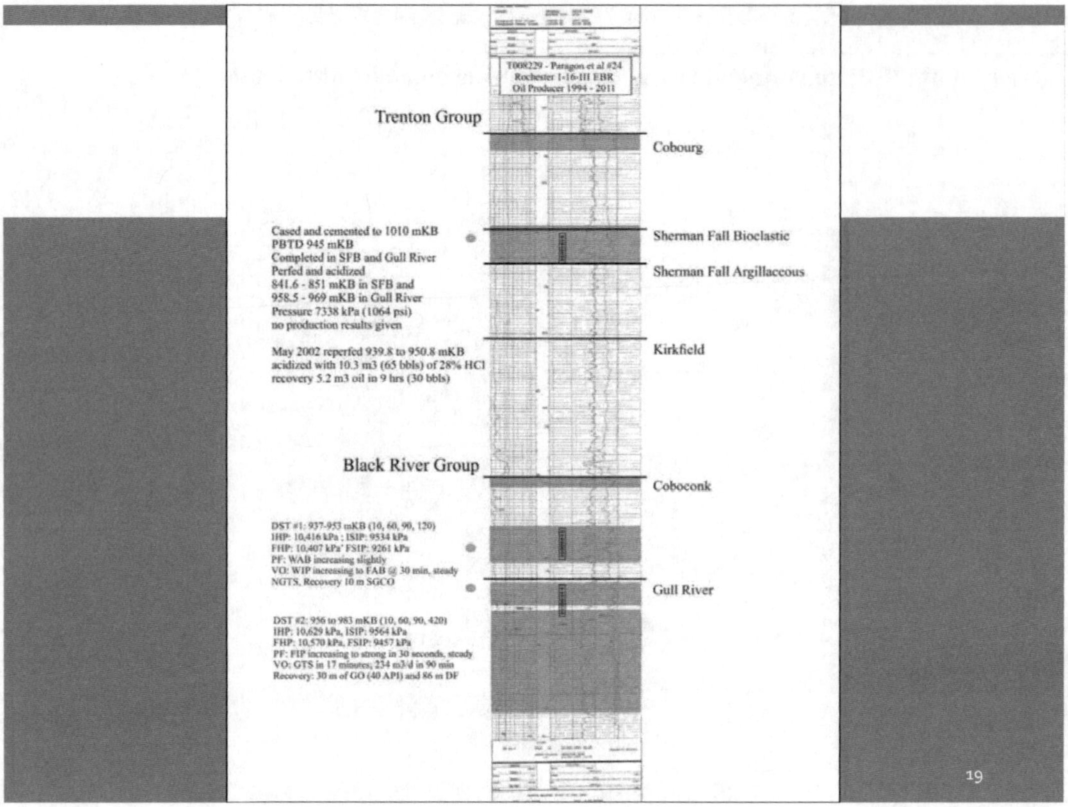
T008229 - Paragon et al #24
Rochester 1-16-III EBR
Oil Producer 1994 - 2011
Trenton Group
Cobourg
Cased and cemented to 1010 mKB
PBTD 945 mKB
Completed in SFB and Gull River
Perfed and acidized
841.6 - 851 mKB in SFB and
958.5 - 969 mKB in Gull River
Pressure 7338 kPa (1064 psi)
no production results given
Sherman Fall Bioclastic
Sherman Fall Argillaceous
May 2002 reperfed 939.8 to 950.8 mKB
acidized with 10.3 m3 (65 bbls) of 28% HCl
recovery 5.2 m3 oil in 9 hrs (30 bbls)
Kirkfield
Black River Group
Coboconk
DST #1: 937-953 mKB (10, 60, 90, 120)
IHP: 10,416 kPa ; ISIP: 9534 kPa
FHP: 10,407 kPa' FSIP: 9261 kPa
PF: WAB increasing slightly
VO: WIP increasing to FAB @ 30 min, steady
NGTS, Recovery 10 m SGCO
Gull River
DST #2: 956 to 983 mKB (10, 60, 90, 420)
IHP: 10,629 kPa, ISIP: 9564 kPa
FHP: 10,570 kPa, FSIP: 9457 kPa
PF: FIP increasing to strong in 30 seconds, steady
VO: GTS in 17 minutes; 234 m3/d in 90 min
Recovery: 30 m of GO (40 API) and 86 m DF
19

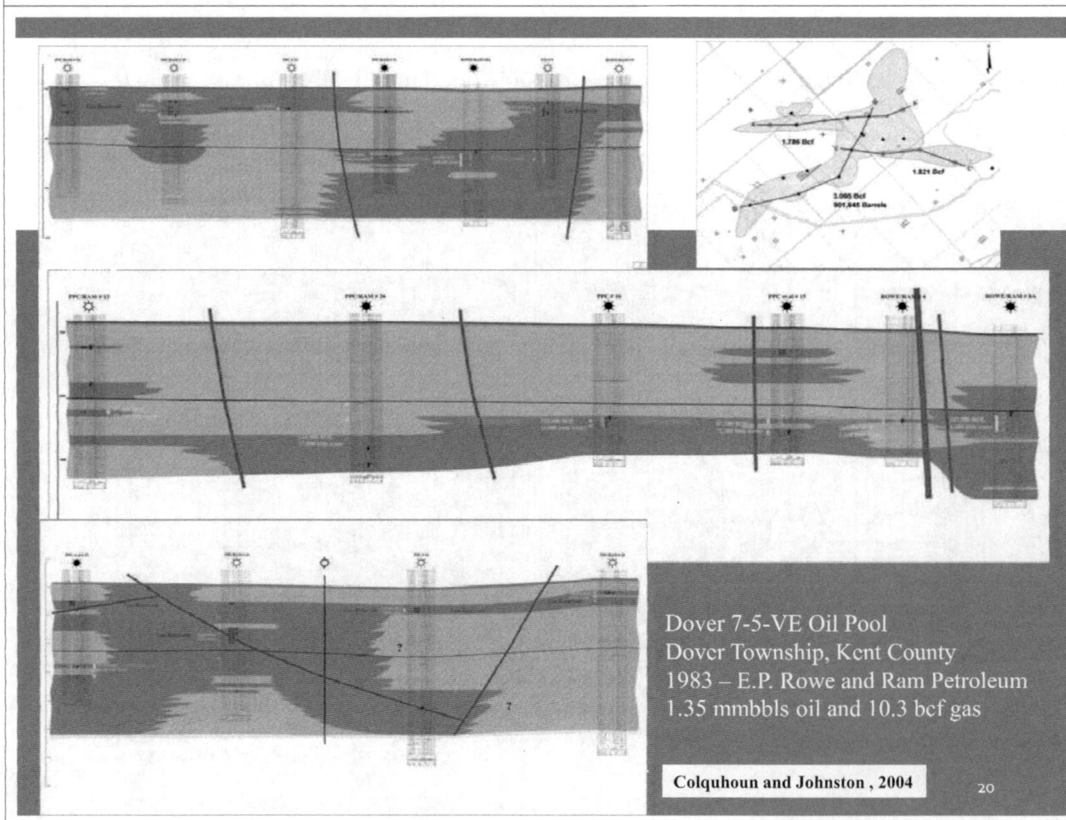
Dover 7-5-VE Oil Pool
Dover Township, Kent County
1983 – E.P. Rowe and Ram Petroleum
1.35 mmbbls oil and 10.3 bcf gas
Colquhoun and Johnston , 2004
20

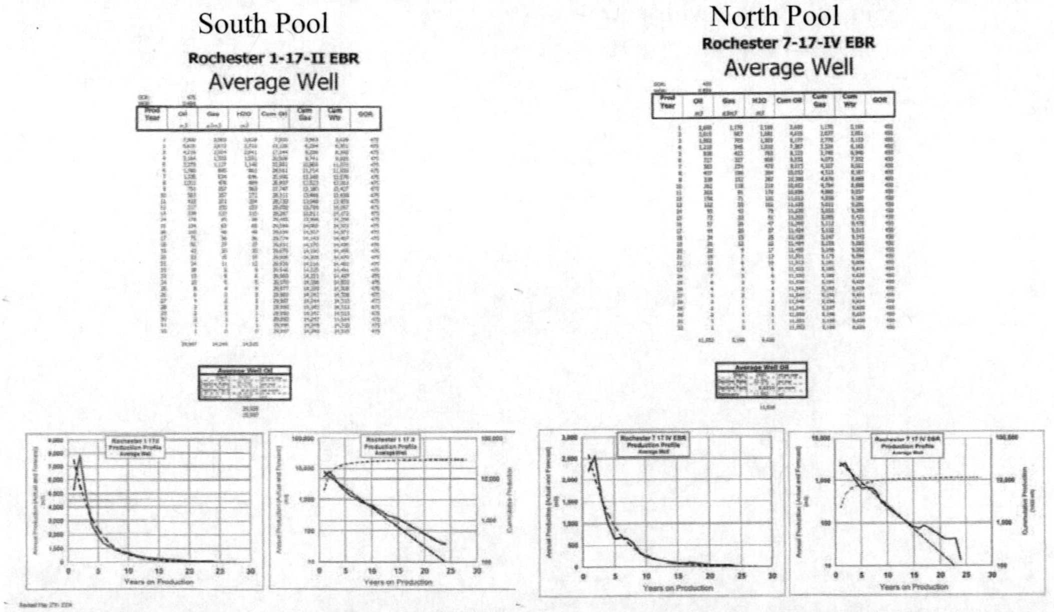

Production profiles for average Trenton-Black River oil wells at Rochester (Golder 2005)
Rochester 1-17-II in Rochester South Oil Pool; Rochester 7-17-IV in Rochester North Oil Pool

21

Cumulative Oil Production (bbls) to 2015

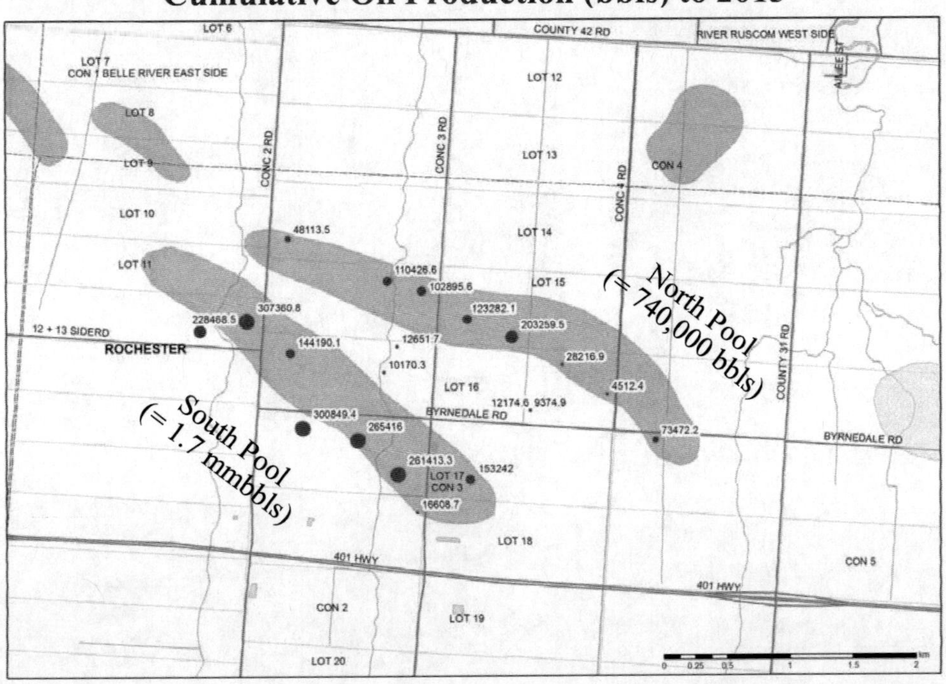

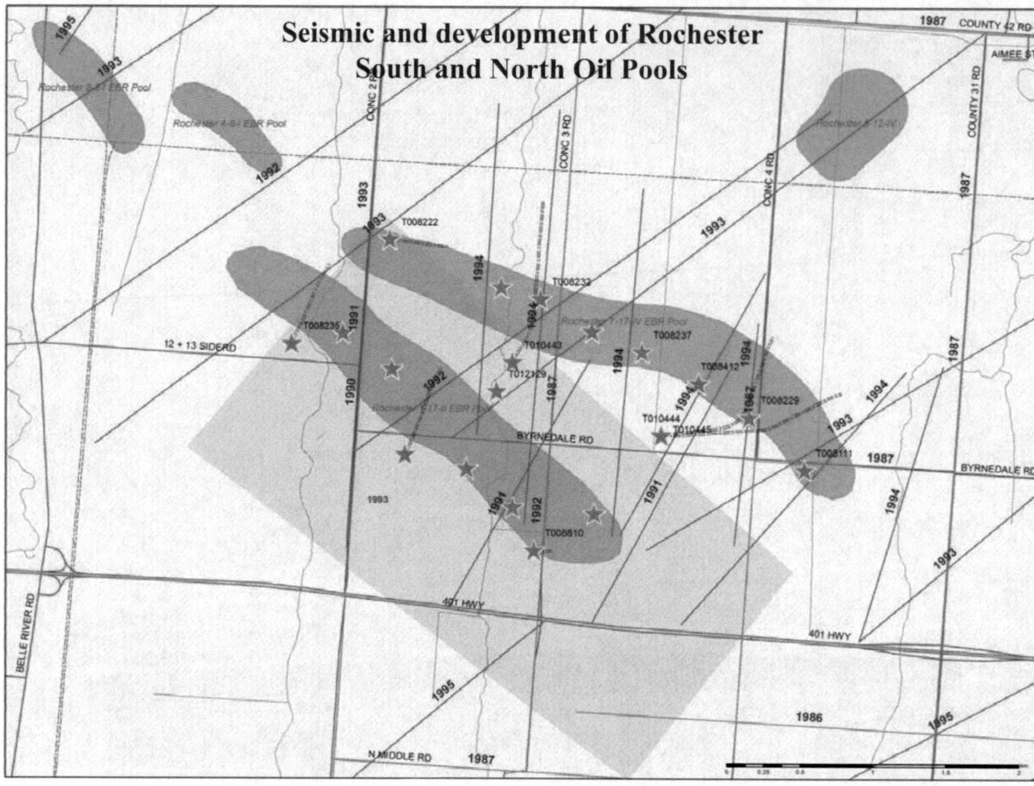
Seismic and development of Rochester
South and North Oil Pools
COUNTY 42 RD
AIMEE ST
CONC 2 RD
CONC 3 RD
CONC 4 RD
COUNTY 31 RD
12 + 13 SIDERD
BYRNEDALE RD
401 HWY
BELLE RIVER RD
N MIDDLE RD
T008222
T008232
T008235
T010443
T012129
T008237
T008412
T008229
T010444
T010445
T008111
T008610
1986
1987
1990
1991
1992
1993
1994
1995

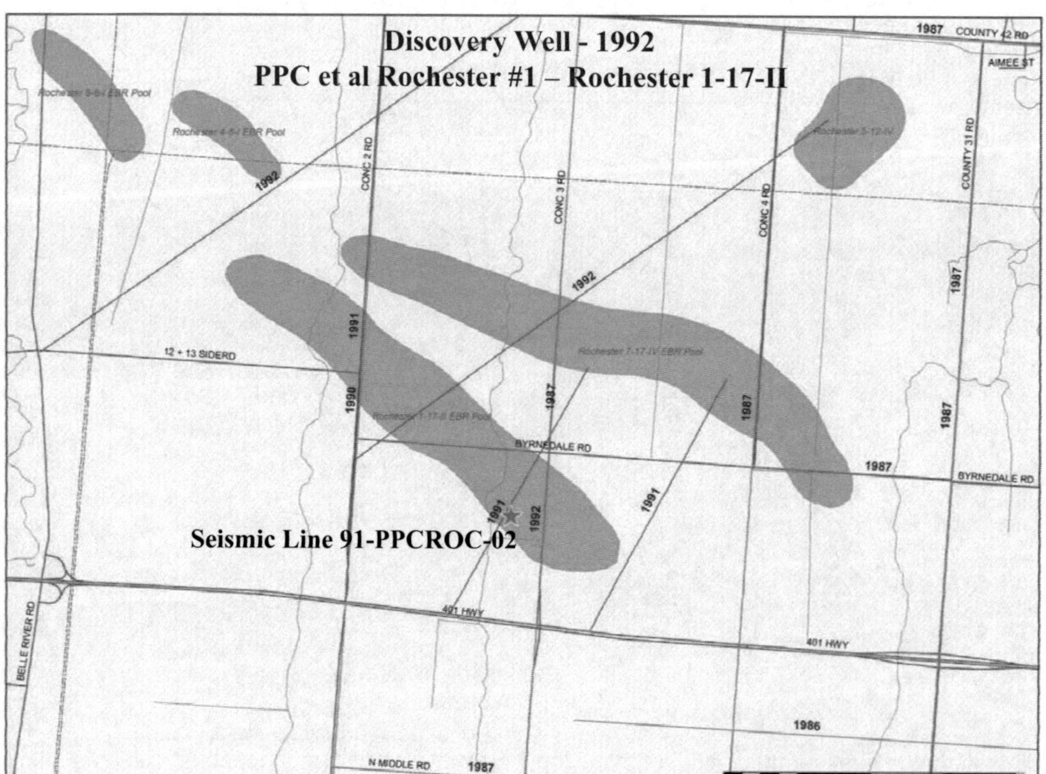
Discovery Well - 1992
PPC et al Rochester #1 – Rochester 1-17-II
Seismic Line 91-PPCROC-02
COUNTY 42 RD
AIMEE ST
CONC 2 RD
CONC 3 RD
CONC 4 RD
COUNTY 31 RD
12 + 13 SIDERD
BYRNEDALE RD
401 HWY
BELLE RIVER RD
N MIDDLE RD
1986
1987
1990
1991
1992

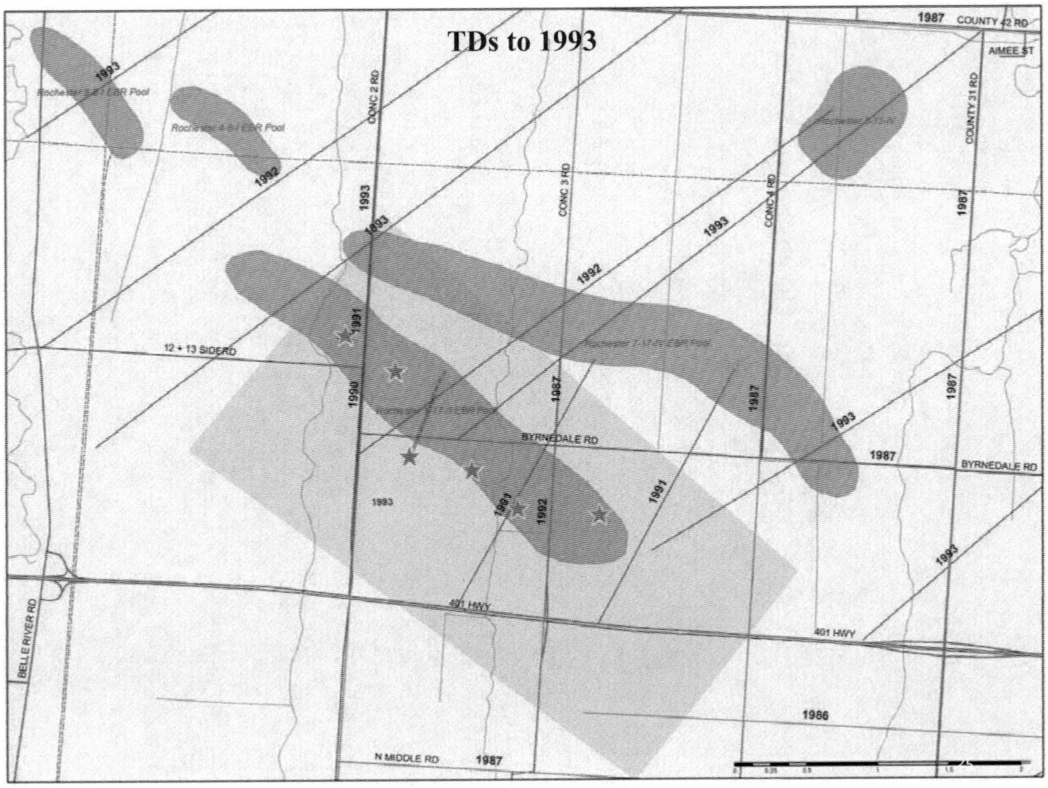
TDs to 1993
1987
COUNTY 42 RD
AIMEE ST
COUNTY 31 RD
CONC 2 RD
CONC 3 RD
CONC 4 RD
12 + 13 SIDERD
BYRNEDALE RD
401 HWY
BELLE RIVER RD
N MIDDLE RD
1986
1990
1991
1992
1993

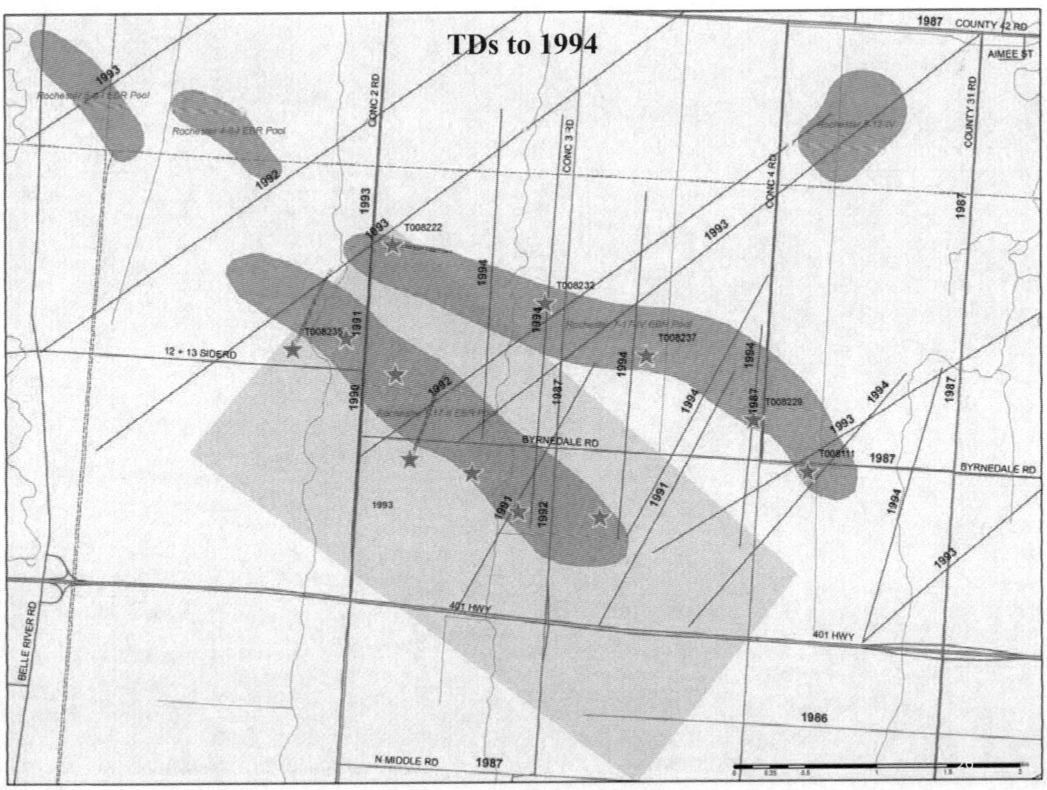
TDs to 1994
1987
COUNTY 42 RD
AIMEE ST
COUNTY 31 RD
CONC 2 RD
CONC 3 RD
CONC 4 RD
12 + 13 SIDERD
BYRNEDALE RD
401 HWY
BELLE RIVER RD
N MIDDLE RD
T008222
T008232
T008237
T008235
T008229
T008111
1986
1990
1991
1992
1993
1994

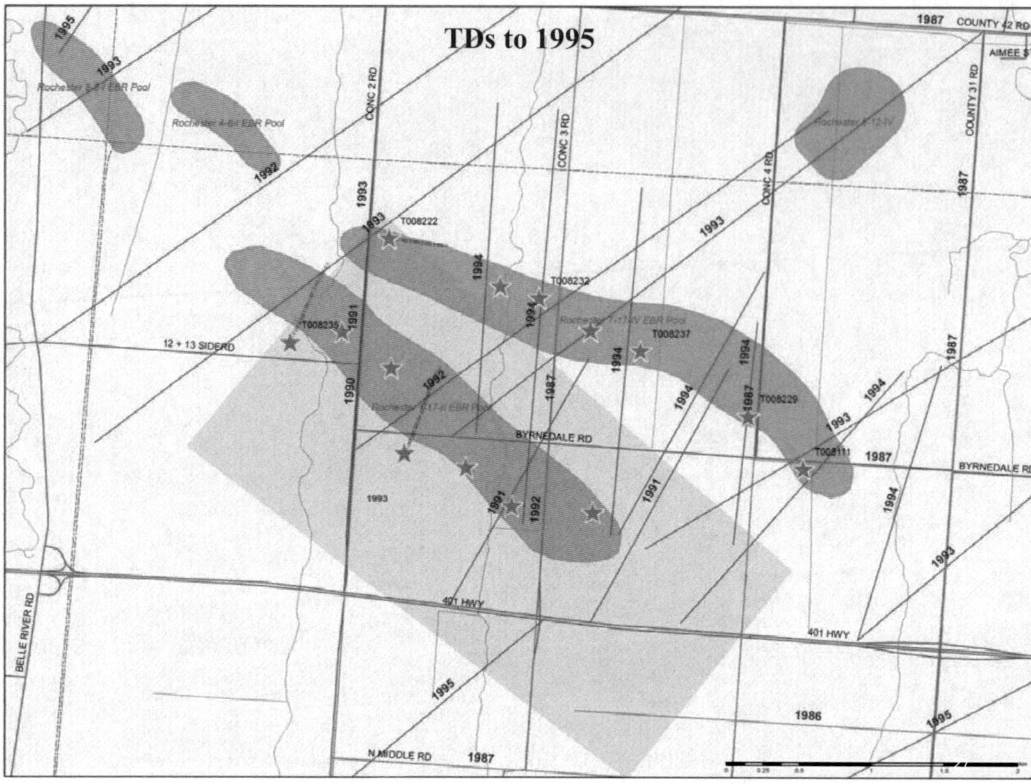
TDs to 1995
1987 COUNTY 42 RD
AIMEE ST
COUNTY 31 RD
CONC 2 RD
CONC 3 RD
CONC 4 RD
Rochester 4-64 EBR Pool
12 + 13 SIDERD
BYRNEDALE RD
401 HWY
BELLE RIVER RD
N.MIDDLE RD
T008222
T008232
T008235
T008237
T008229
T008111

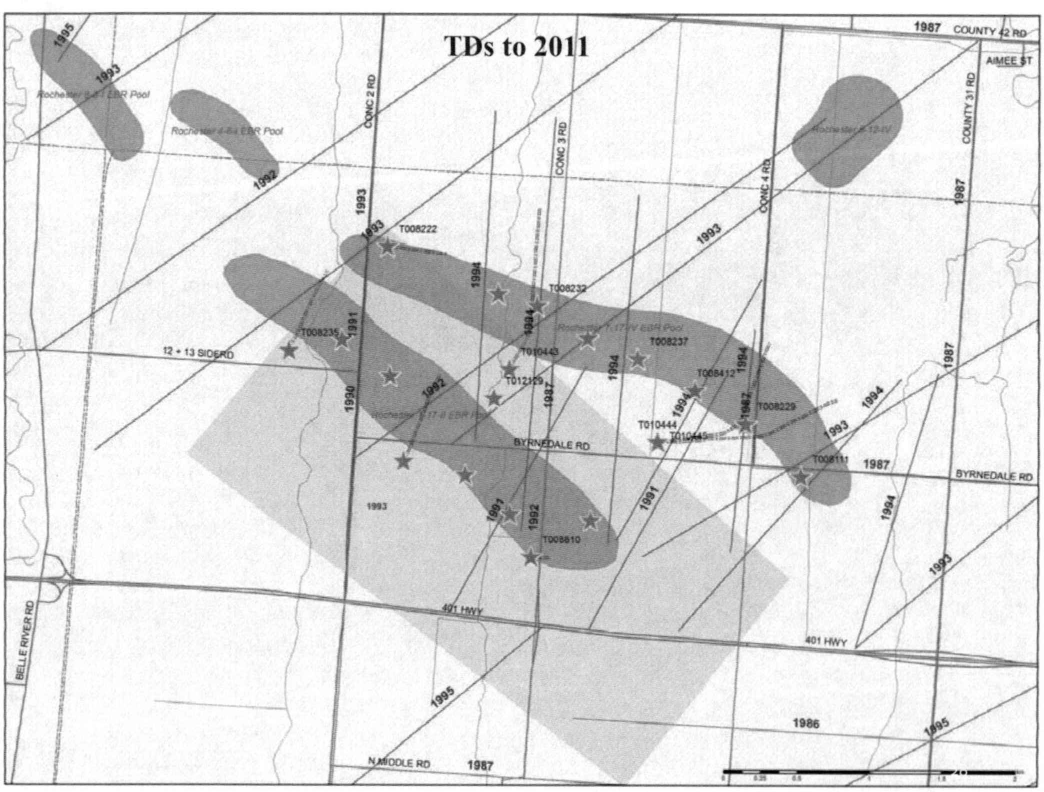
TDs to 2011
1987 COUNTY 42 RD
AIMEE ST
COUNTY 31 RD
CONC 2 RD
CONC 3 RD
CONC 4 RD
Rochester 4-64 EBR Pool
12 + 13 SIDERD
BYRNEDALE RD
401 HWY
BELLE RIVER RD
N.MIDDLE RD
T008222
T008232
T008235
T010443
T012129
T008237
T008412
T008229
T010444
T010445
T008111
T008810

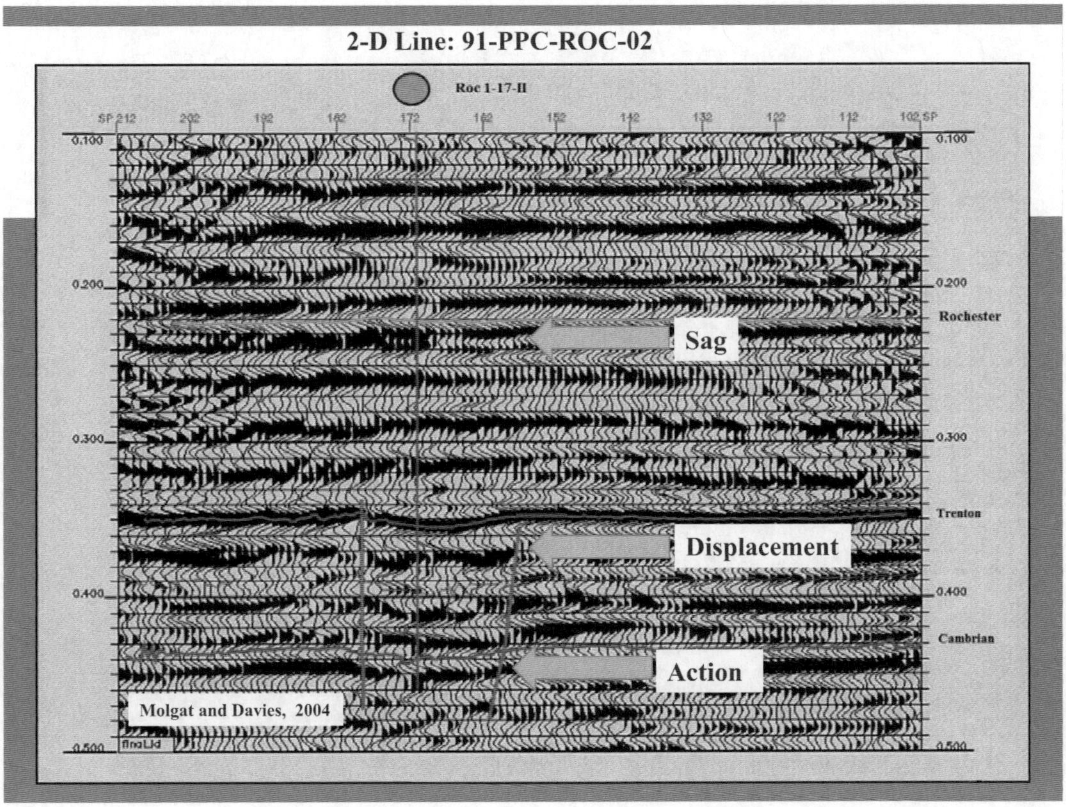

2-D Line: 91-PPC-ROC-02
Roc 1-17-II
Rochester
Sag
Trenton
Displacement
Cambrian
Action
Molgat and Davies, 2004

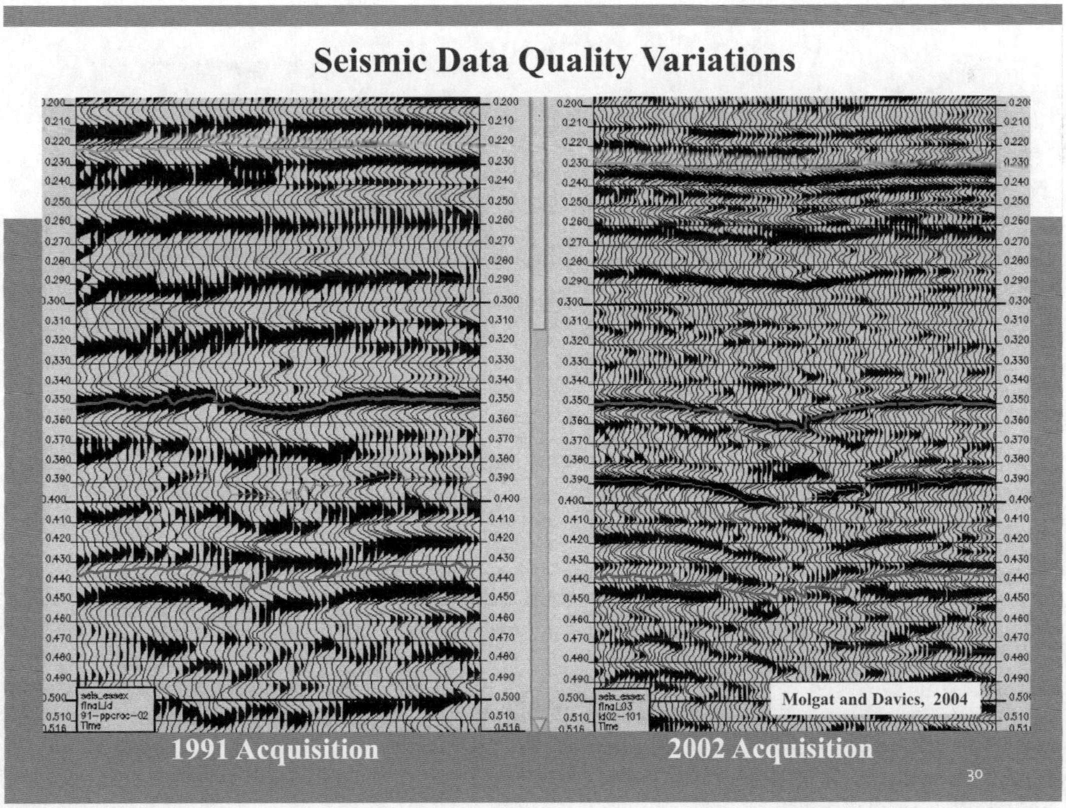

Seismic Data Quality Variations
Molgat and Davies, 2004
1991 Acquisition
2002 Acquisition
30

Sample - Trenton Seismic Section

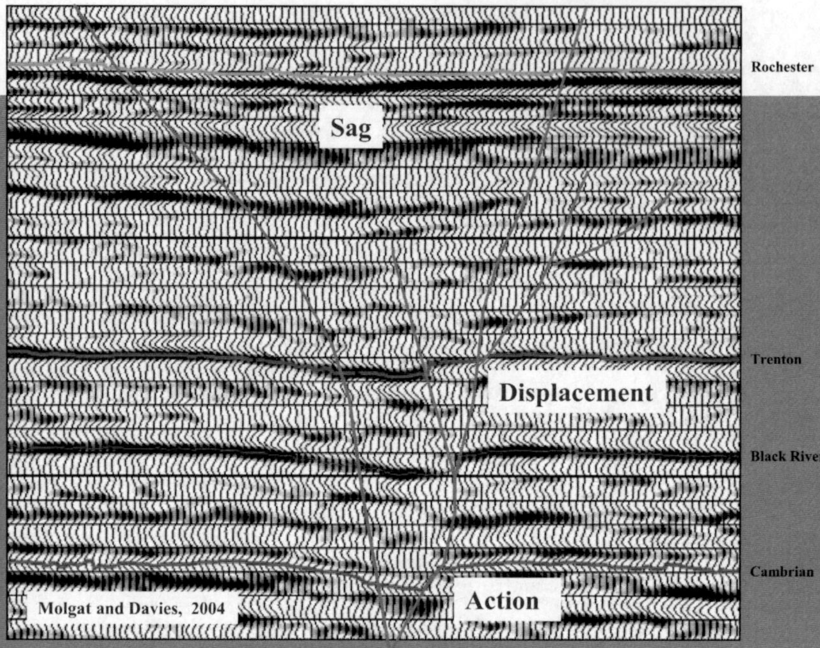

31

Trenton Time Structure

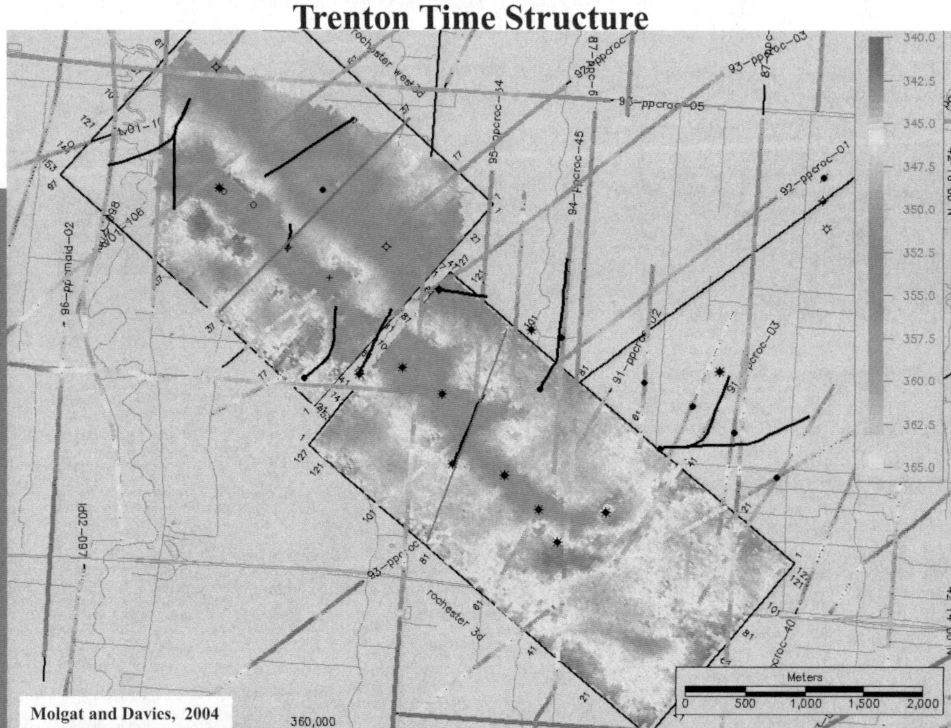

32

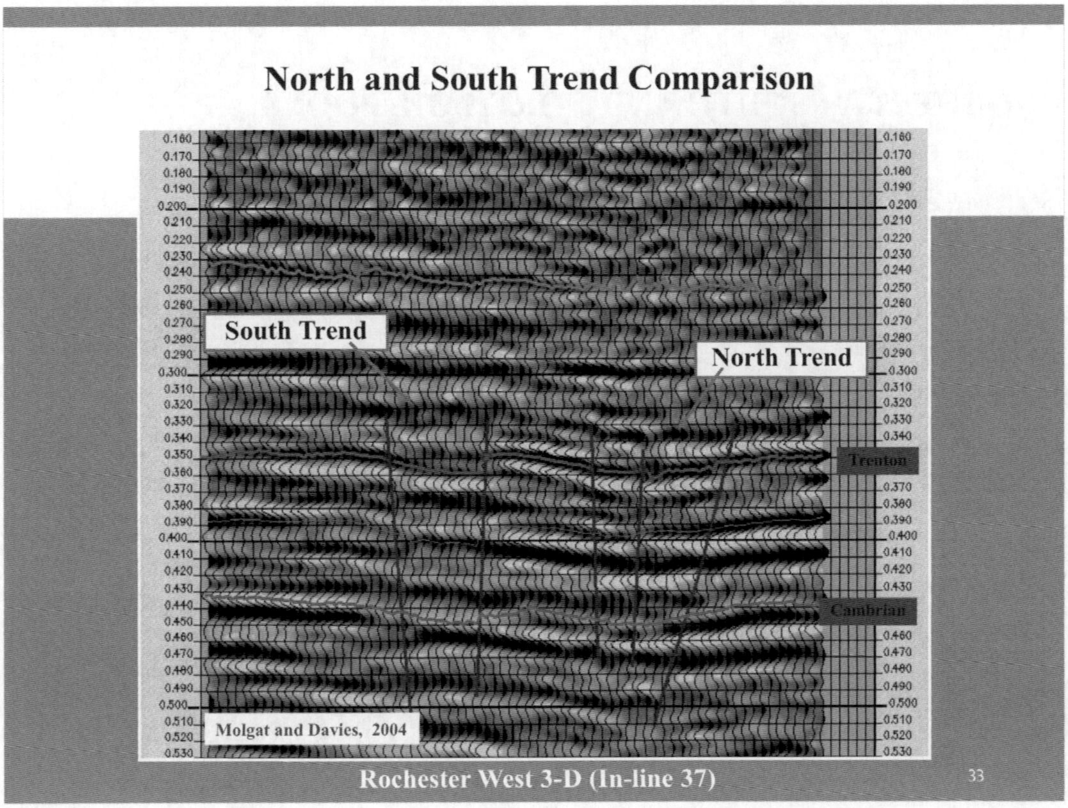
North and South Trend Comparison
South Trend
North Trend
Trenton
Cambrian
Molgat and Davies, 2004
Rochester West 3-D (In-line 37)
33

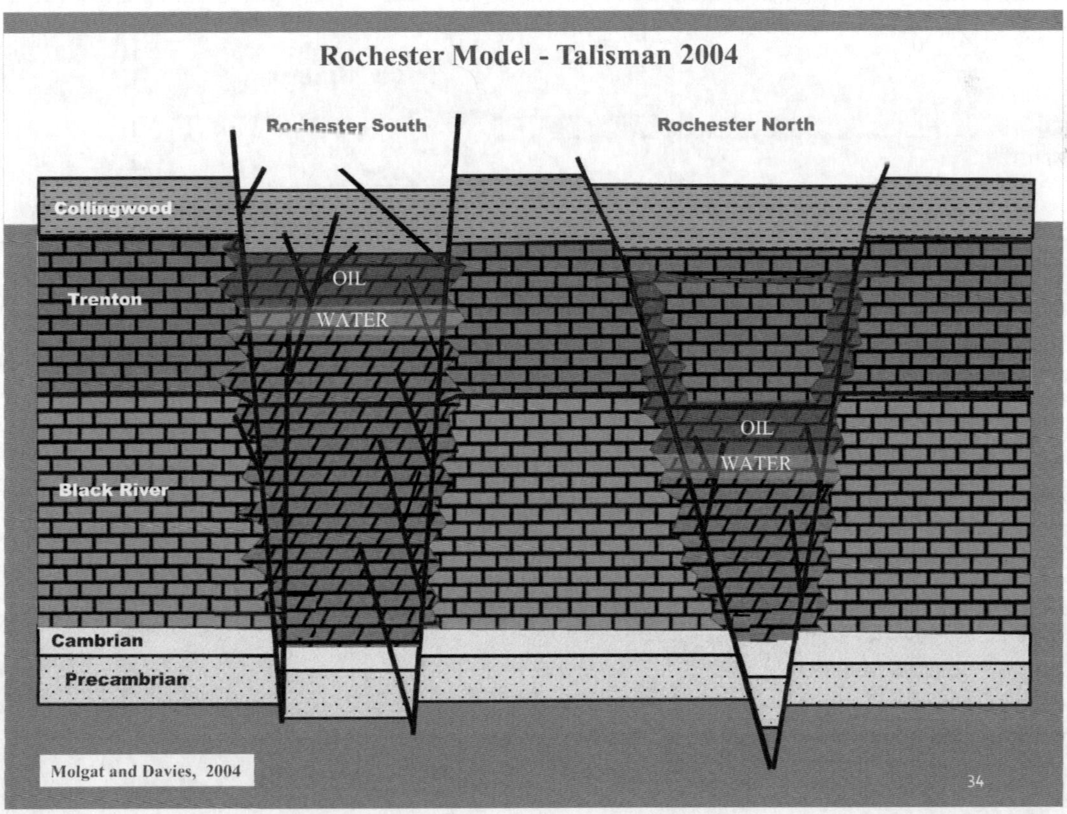
Rochester Model - Talisman 2004
Rochester South
Rochester North
Collingwood
Trenton
OIL
WATER
Black River
OIL
WATER
Cambrian
Precambrian
Molgat and Davies, 2004
34

ROCHESTER SOUTH AND NORTH OIL POOLS

•Rochester South Oil Pool was discovered in 1992 by Paragon Petroleum Corporation

•Rochester North Oil Pool was discovered in 1994 by Paragon Petroleum Corporation

•Hydrothermally dolomitized (HTD) Trenton - Black River Group carbonates

•Both pools trend NW-SE, are up to 5 km long and are 300-600 m wide

•Controlled by faults and fractures created by strike-slip movements

•Rochester South and North Oil Pools have 26 producing wells

•Both trends produce oil from the Trenton but the north trend produces oil from the Black River

•Reservoirs: 840 m (2760') and 940 m (3080'); 7800 kPa (1130 psi) and 25°C

•End 2015, cumulative oil production was 2.5 mmbbls

•Estimated to be >92% of the recoverable reserves using a 15% recovery factor

•Porosity: 3-15% (8%) matrix, 18-45% fractures; Permeability: 10–300 md (150 md) in matrix and 2-10 Darcies for fractures

•Oil is sweet, 40° API, with solution gas; water saturation 15 to 40%, irreducible 15 to 20%

•Typical decline rates between 15 and 25% per year

35

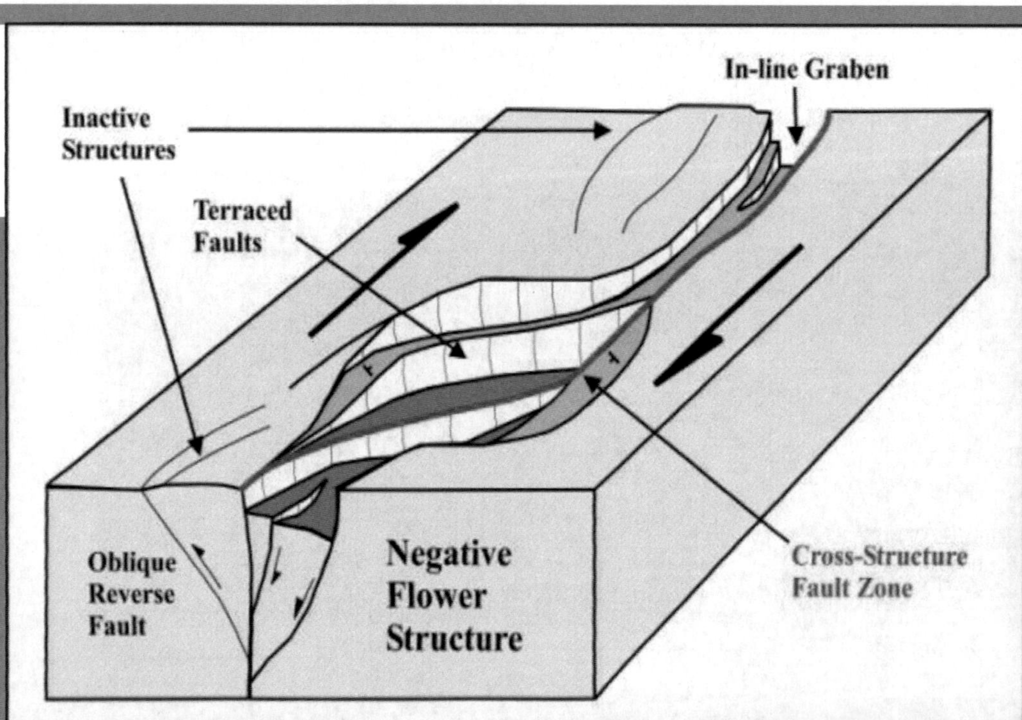

Isometric representation of a pull-apart basin (modified from Dooley and McClay, 1997) showing sag or "graben" structure.

36

Fracture pattern on an anticlinal form

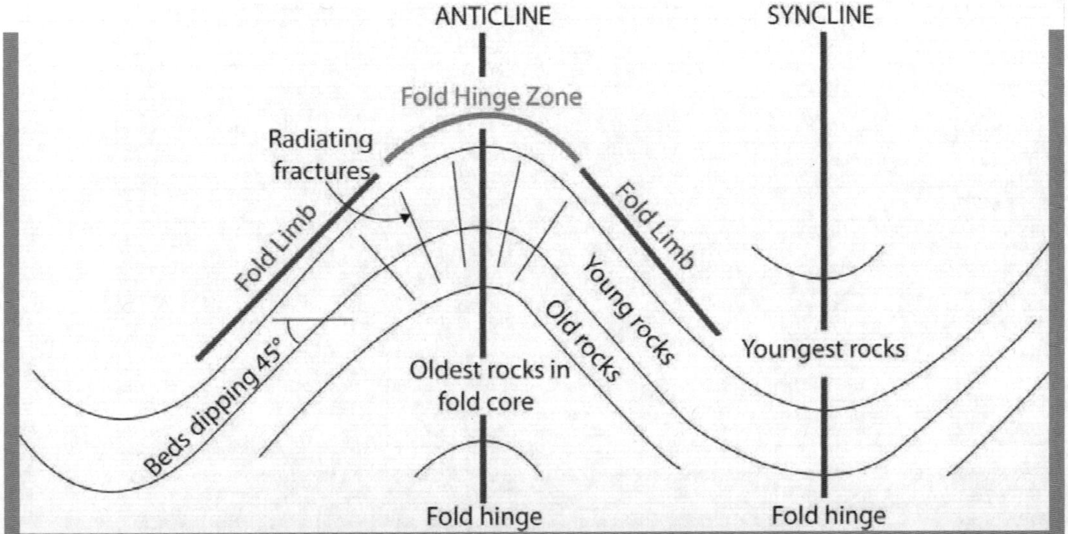

Fracture pattern and geometry over an Anticlinal form; near vertical on top and high angle on sides

37

Conceptual model of a Trenton-Black River reservoir at Rochester North and oil and gas pools at Dover. Modified from Colquhoun and Johnston (2004).

38

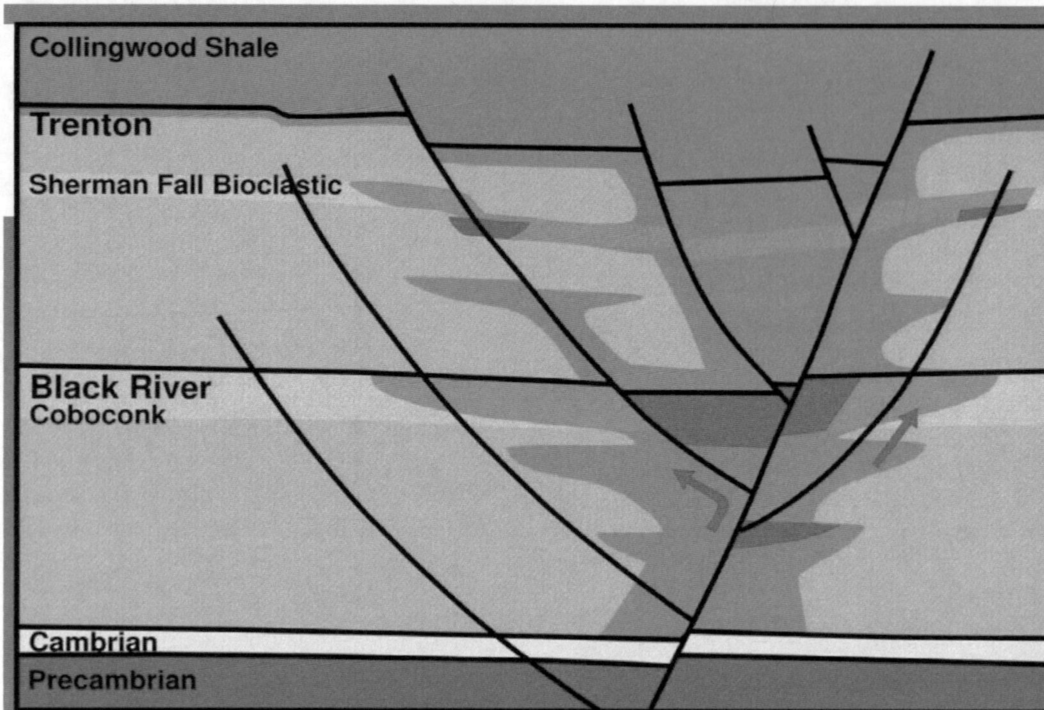

Conceptual model of a Trenton-Black River reservoir at Rochester South and other oil pools located in Essex and Kent Counties. Modified from Colquhoun (2012).

39

EXPLORATION AND DEVELOPMENT HISTORY AND FUTURE POTENTIAL

- Play has been active since the late 1800's, last discovery in 1992 and 1994 at Rochester
- Hydrothermally dolomitized (HTD) Trenton - Black River Group limestones
- Pools trend NW-SE and E-W, up to 15 km long and 300-1000 m wide (Goldsmith-Lakeshore)
- Oil is sweet, 40 – 42° API, with solution gas; typical decline rates between 15 and 25% per year
- Reservoir: 650 m (2130') to 1050 m (3450'); 7800 to 8050 kPa or 1130 to 1170 psi at 25°C;
- Net pay - up to 10 m or 30 feet; Porosity: 3-15% (8%) matrix, 18-45% fractures; Permeability: 10–300 md (150 md) in matrix and 2-10 Darcies for fractures
- Source rock – Middle Ordovician carbonates and Upper Ordovician shale; carbonaceous
- Seal – upper cap dolomite and overlaying shale; lateral seal is impermeable limestones
- Play occupies approximately 120,000 km^2
- Potential oil recoverable: 40 mmbbls; 23 mmbbls recoverable; 17 mmbbls remaining (Golder 2005)
- Potential gas recoverable: 281 bcf; 41 bcf recoverable; 240 bcf remaining (Golder 2005)
- Oil production in Essex and Kent Counties; oil potential in northwest Lambton County where Cambrian sandstones are thick; gas production from fractured limestones in Lambton County over Arch
- Gas prone from Elgin County east to Welland County and north over the Algonquin Arch
- Historical success rates: 80's and 90's -- 67% for exploration wells and 80% for development wells
- Seismically driven play, 2-D to locate test drill target, 3-D to explore and develop pool(s)

40

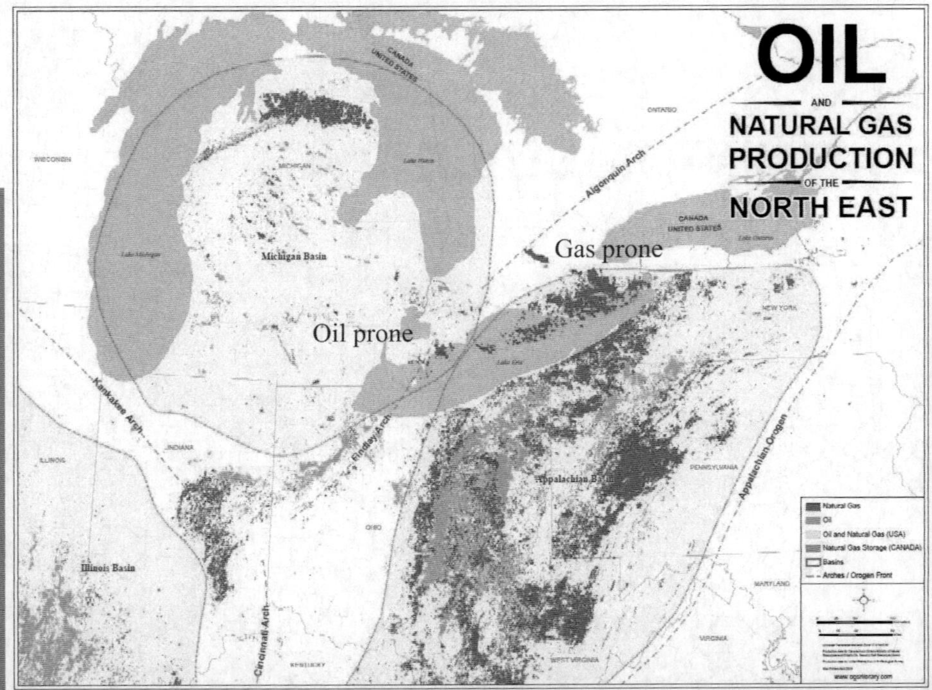

Map showing Ordovician oil and gas pools and future exploration potential

41

THANK YOU FOR LISTENING

ANY QUESTIONS?

42

TRENTON-BLACK RIVER HYDROTHERMAL RESERVOIRS OF THE MICHIGAN BASIN

Bill Van Sickel – Michigan

West Bay Exploration company

Stratigraphic Column - Michigan

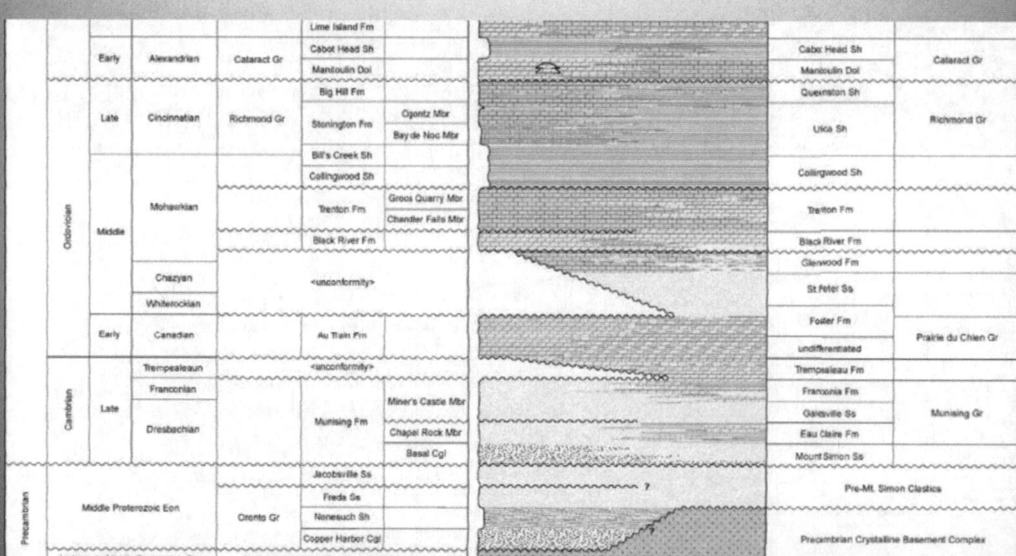

Stratigraphic Nomenclature for Michigan, 2000

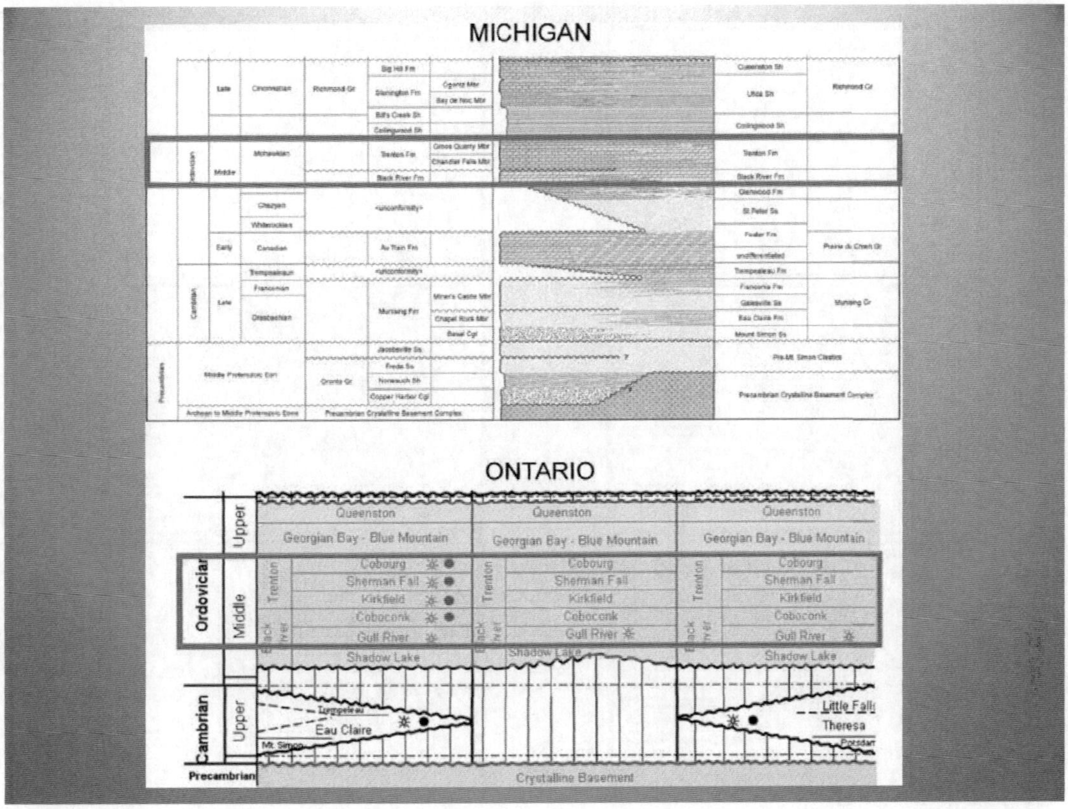
MICHIGAN
ONTARIO
Queenston
Georgian Bay - Blue Mountain
Cobourg
Sherman Fall
Kirkfield
Coboconk
Gull River
Shadow Lake
Trenton
Eau Claire
Mt. Simon
Little Falls
Theresa
Crystalline Basement
Upper
Middle
Ordovician
Cambrian
Precambrian

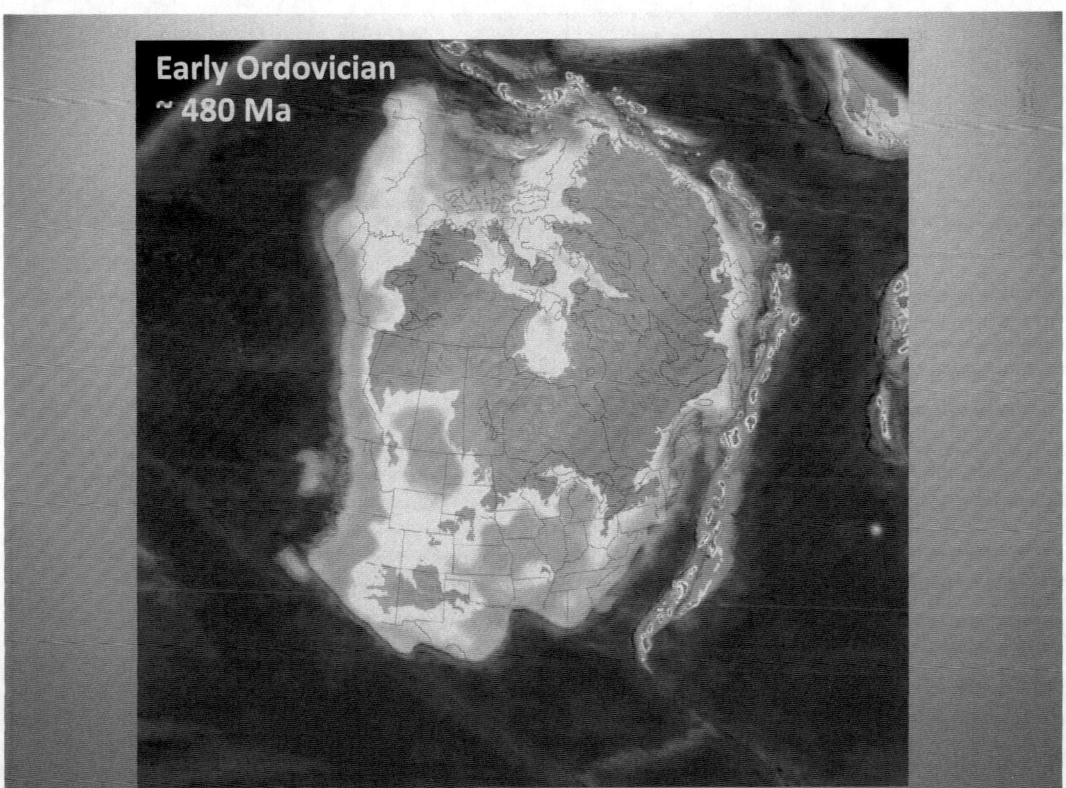
Early Ordovician
~ 480 Ma

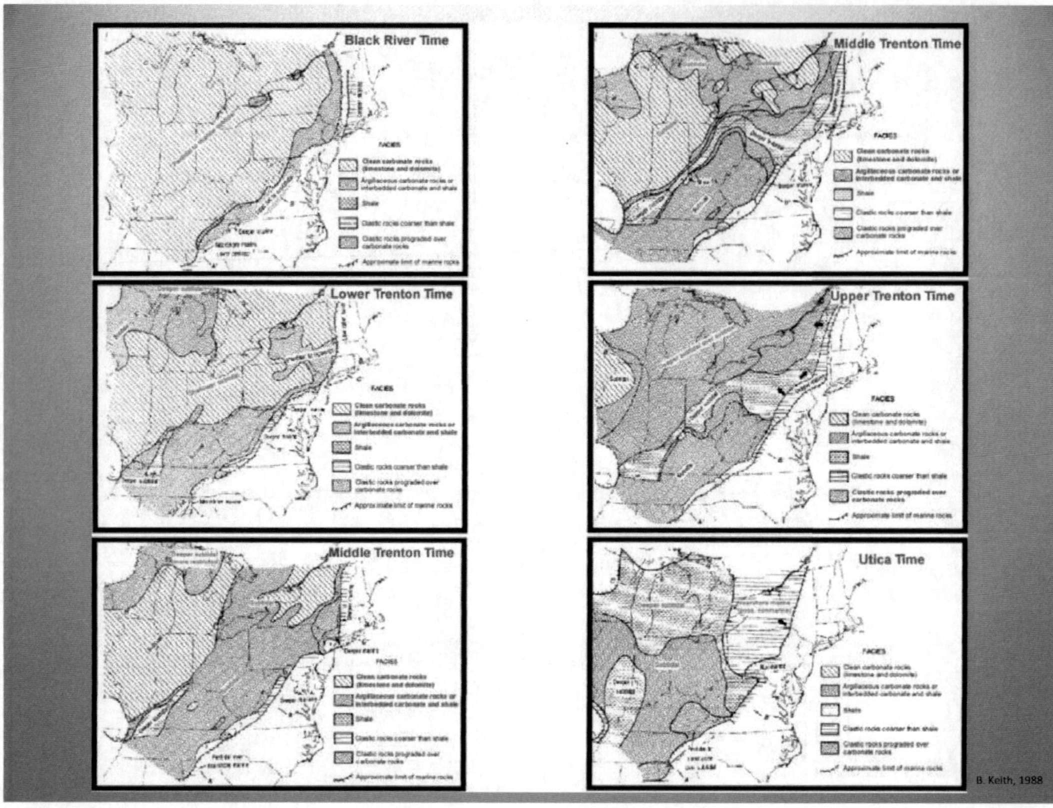

Regional Trenton-Black River S.E. Michigan

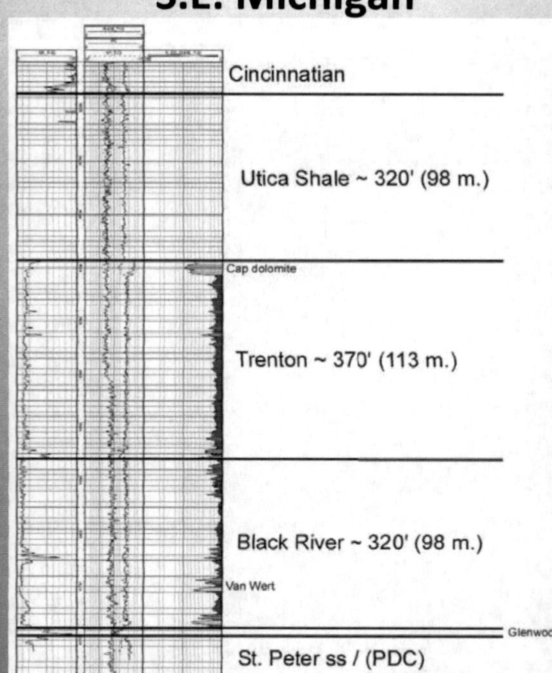

TRENTON-BLACK RIVER RESERVIOR

- Hydrothermal dolomite
- Seal – Utica shale (vertical), Tight limestone (lateral)
- Formed by transtensional stress on existing basement faults during Taconic Orogeny
- Chaotic mix of fractured dolomite/limestone with varying degrees of vugular and matrix porosity
- Solution Gas Drive
- ***EVERY WELL IS DIFFERENT***

Transtensional Stress on existing basement faults during Taconic Orogeny

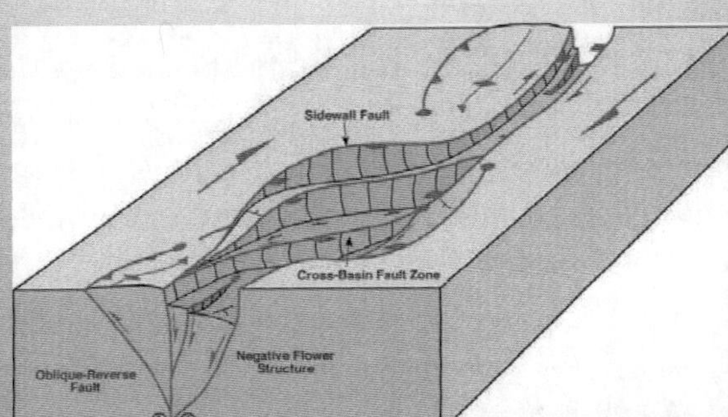

Linear series of extensional, left-lateral wrench faults

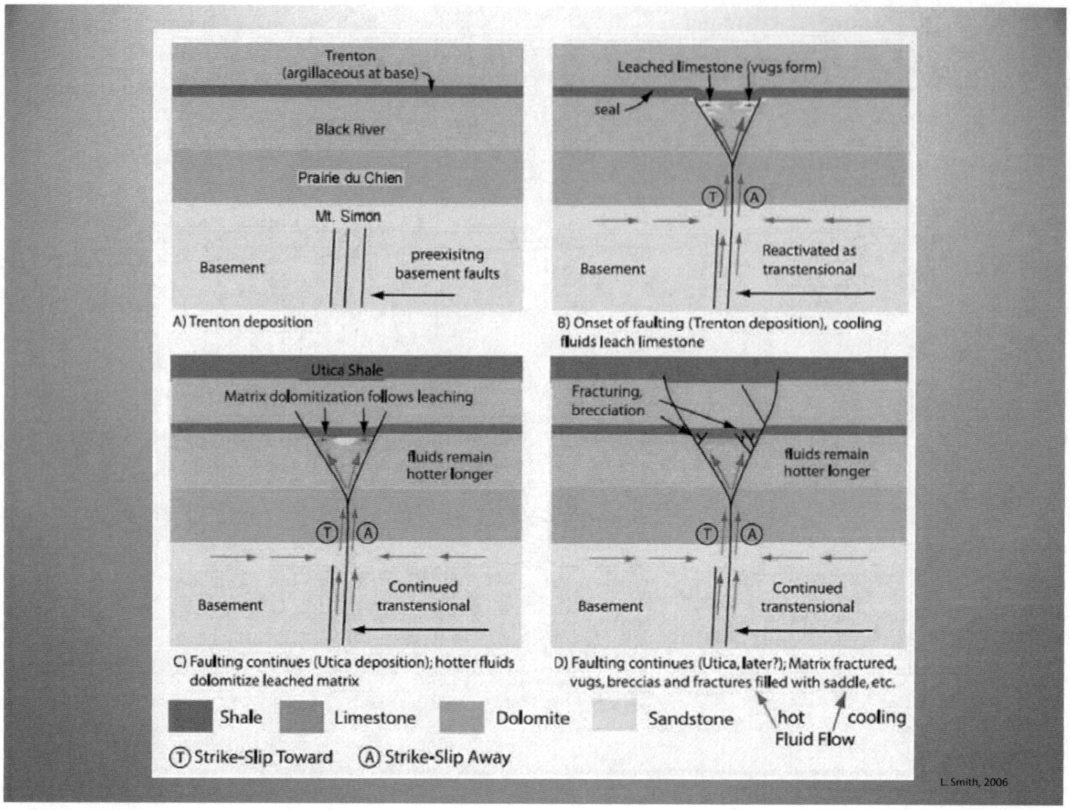

Hydrothermal Fluid Migration

- **Flow along path of least resistance**
- **Controlling Factors : Faults, fractures, shale barriers, bedding planes & facies**
- **These controls on porosity/permeability dictate where the hydrothermal fluids flow in both a vertical & horizontal directions.**

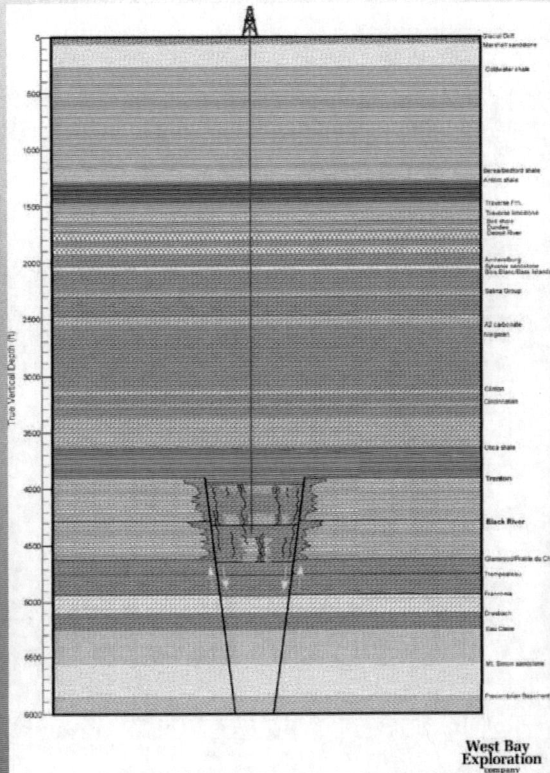

Palatine Bridge Dolomite Outcrop, Mohawk Valley, New York

CONFIDENTIAL

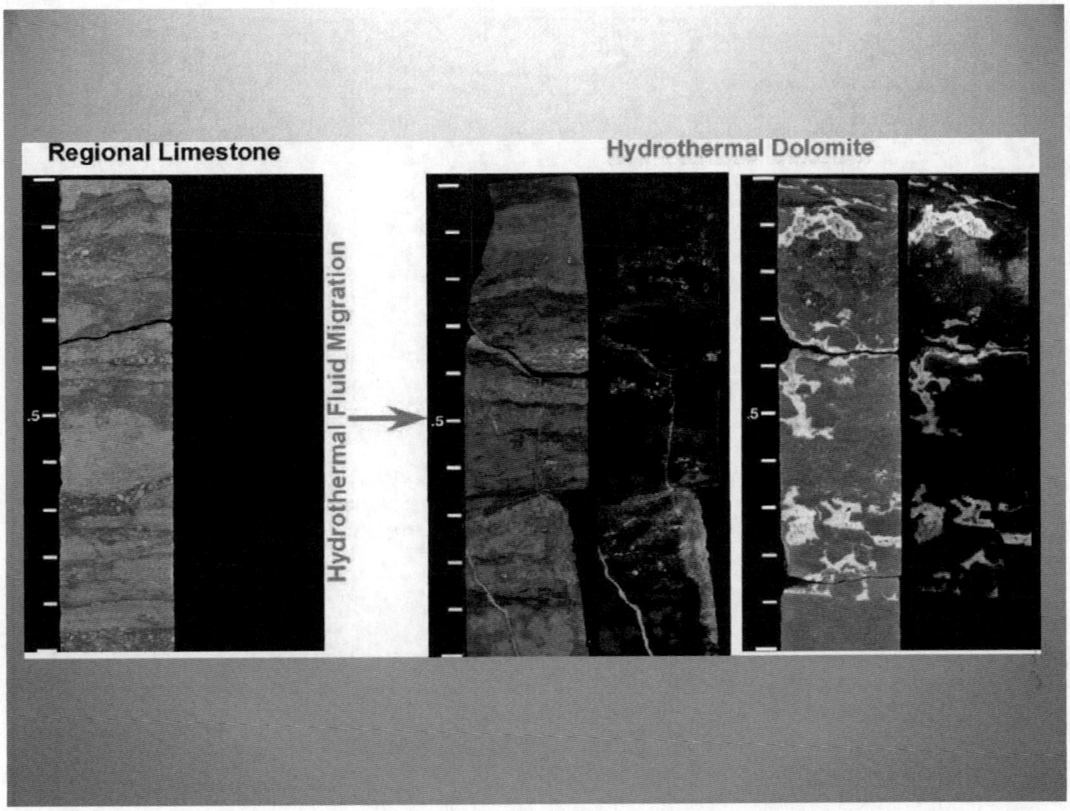
Regional Limestone
Hydrothermal Dolomite
Hydrothermal Fluid Migration
.5
.5
.5

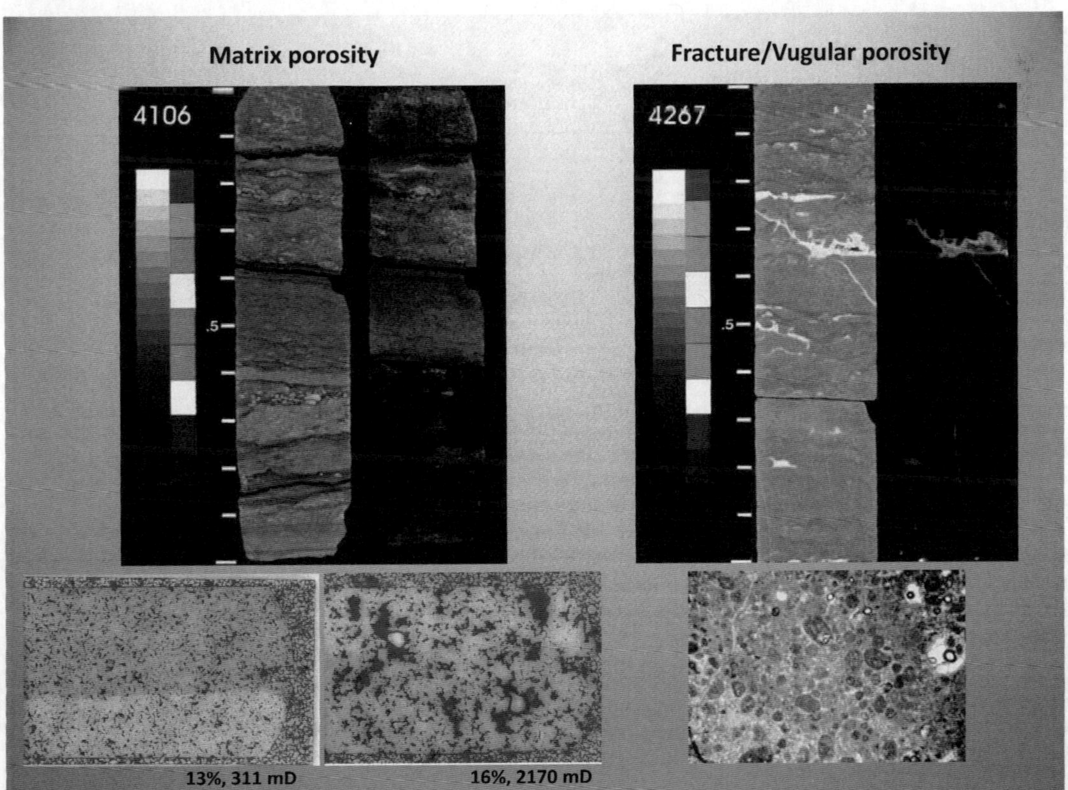
Matrix porosity
Fracture/Vugular porosity
4106
4267
.5
.5
13%, 311 mD
16%, 2170 mD

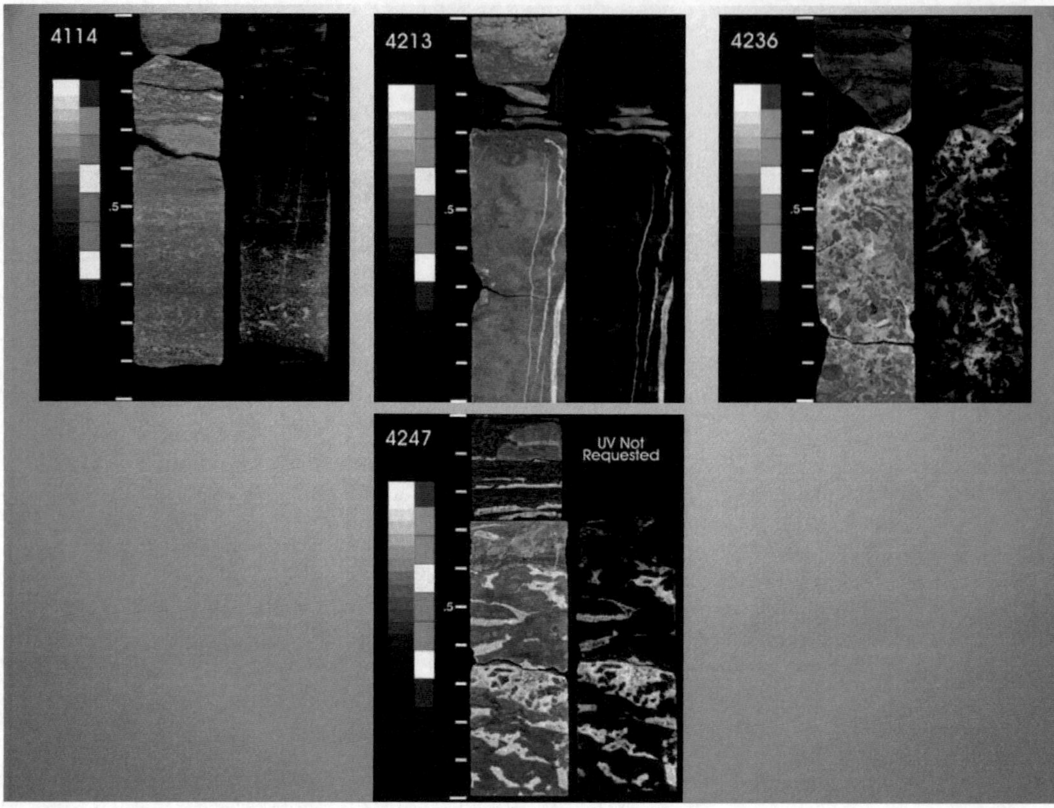
4114
.5
4213
.5
4236
.5
4247
UV Not
Requested
.5

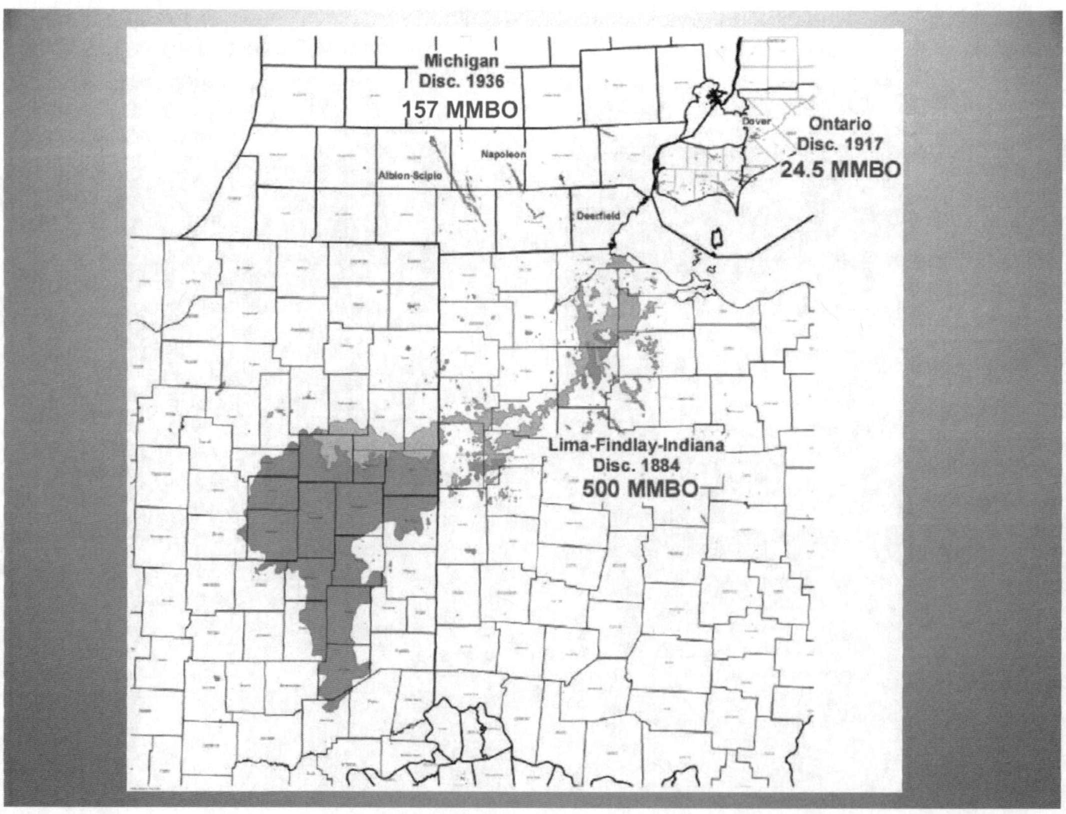
Michigan
Disc. 1936
157 MMBO
Ontario
Disc. 1917
24.5 MMBO
Dover
Napoleon
Albion-Scipio
Deerfield
Lima-Findlay-Indiana
Disc. 1884
500 MMBO

MICHIGAN

TRENTON-BLACK RIVER OIL & GAS FIELDS

TRENTON-BLACK RIVER HISTORIC FIELDS S.E. MICHIGAN

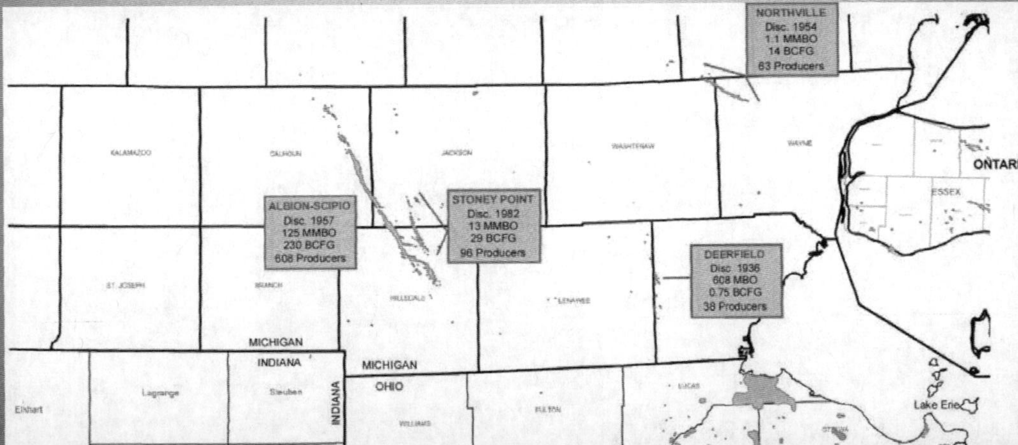

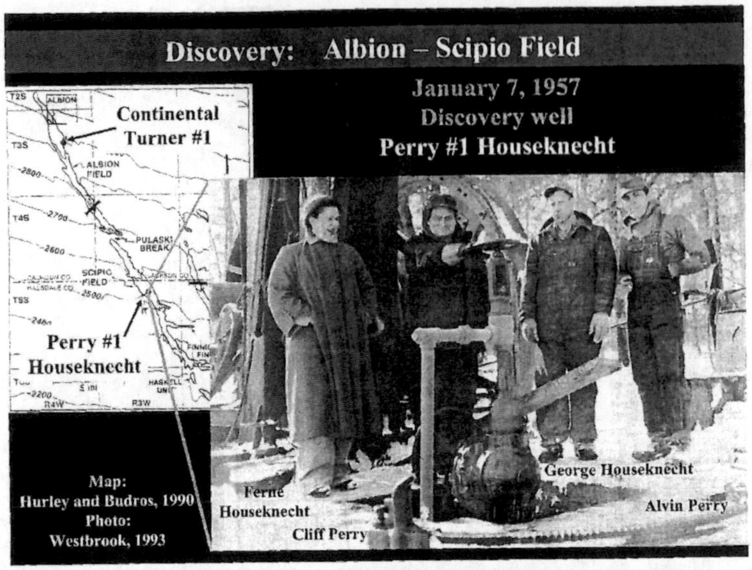

1955 – Houseknecht #1
Originally drilled for Devonian gas – DRY
1957 – Deepened by advice of psychic friend
Encountered oil @ 3900′ – 140 BOPD

TRENTON-BLACK RIVER
RESURGENCE

RICE CREEK DISCOVERY
2006

West Bay Exploration Co.

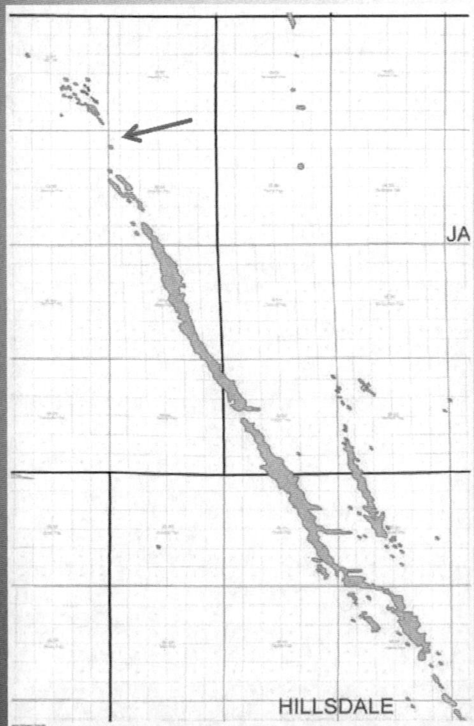

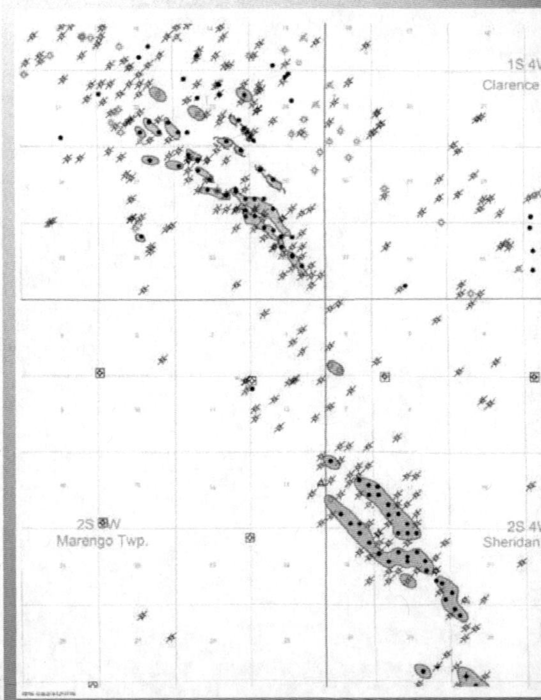

West Bay programmed and shot a 7 ½ mi^2 3-D seismic survey

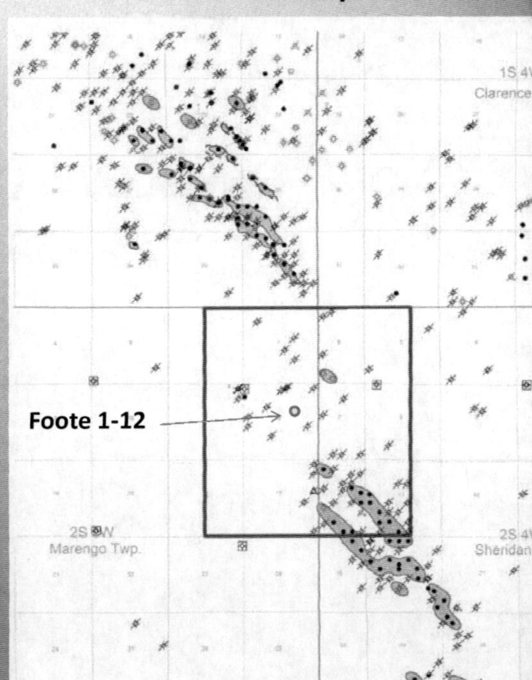

Cum. 3,033,477 BO

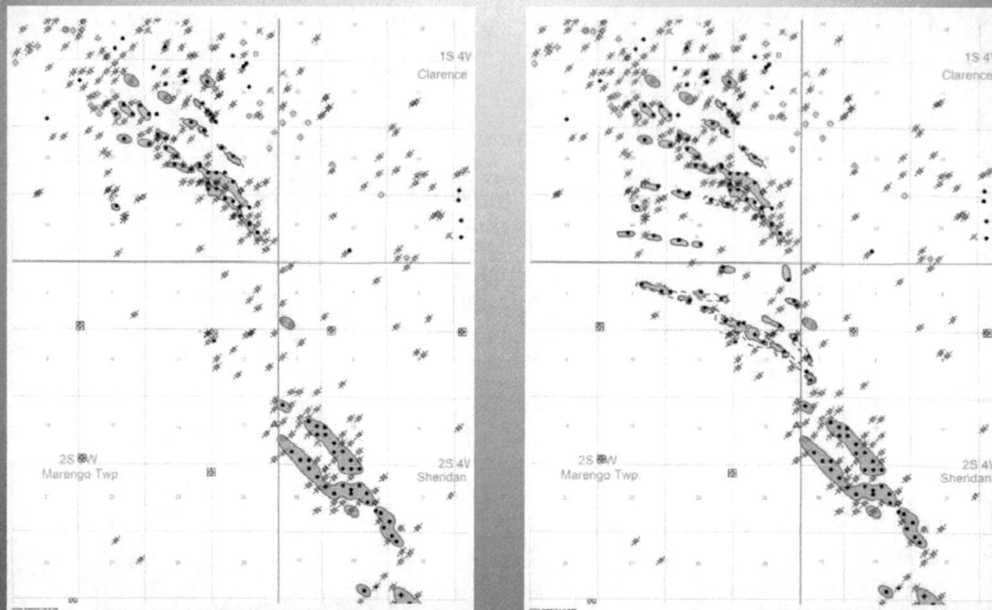

NAPOLEON DISCOVERY 2008

West Bay Exploration Co.

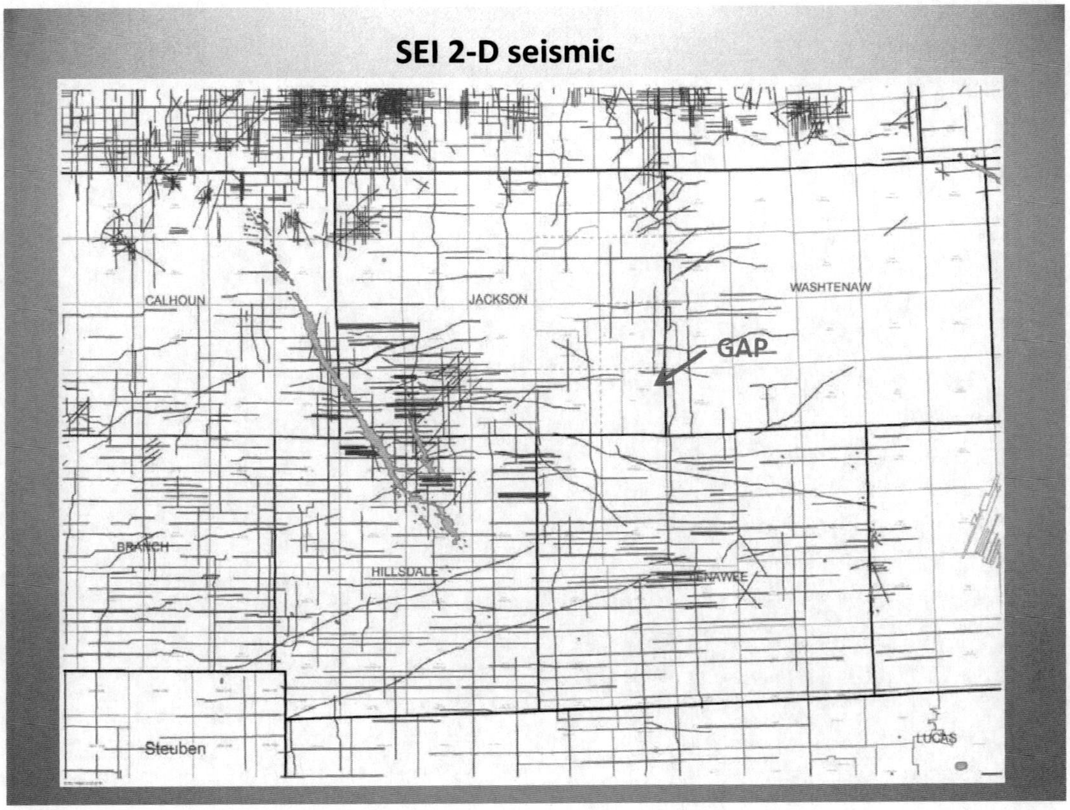
SEI 2-D seismic
CALHOUN
JACKSON
WASHTENAW
GAP
BRANCH
HILLSDALE
Steuben
LUCAS

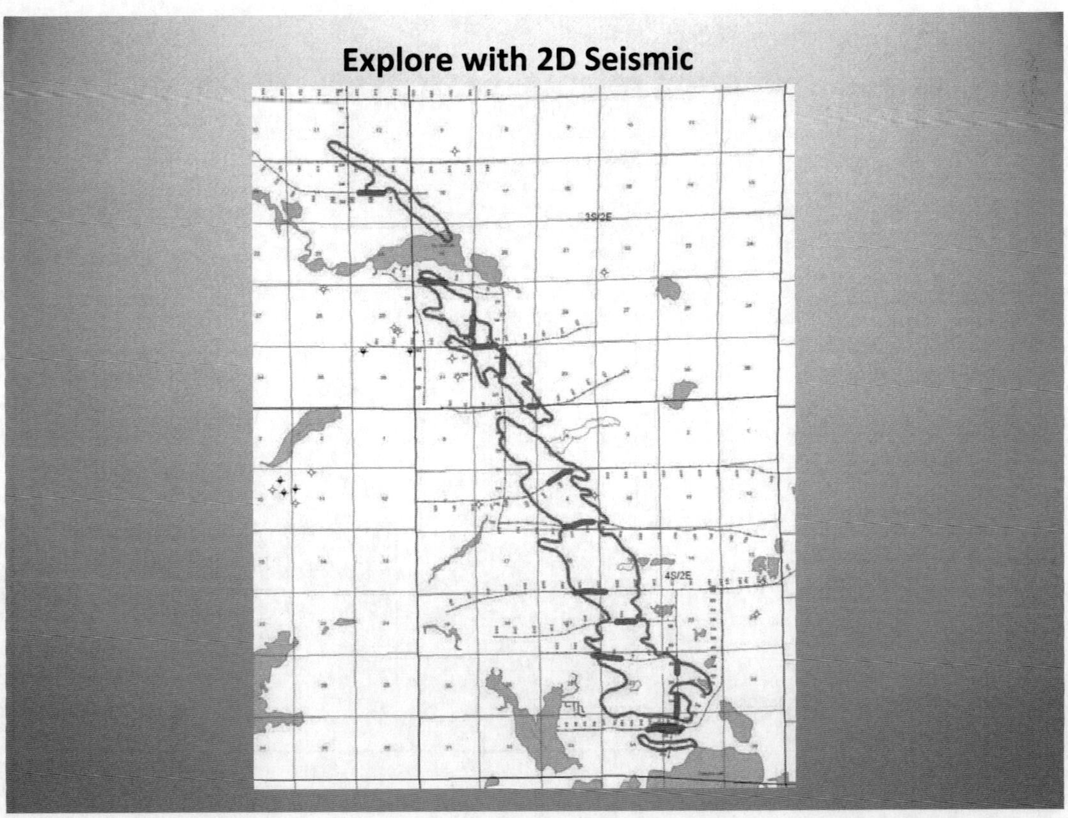
Explore with 2D Seismic
3S/2E
4S/2E

CONFIDENTIAL

Facies distribution relative to G/O & O/W contacts

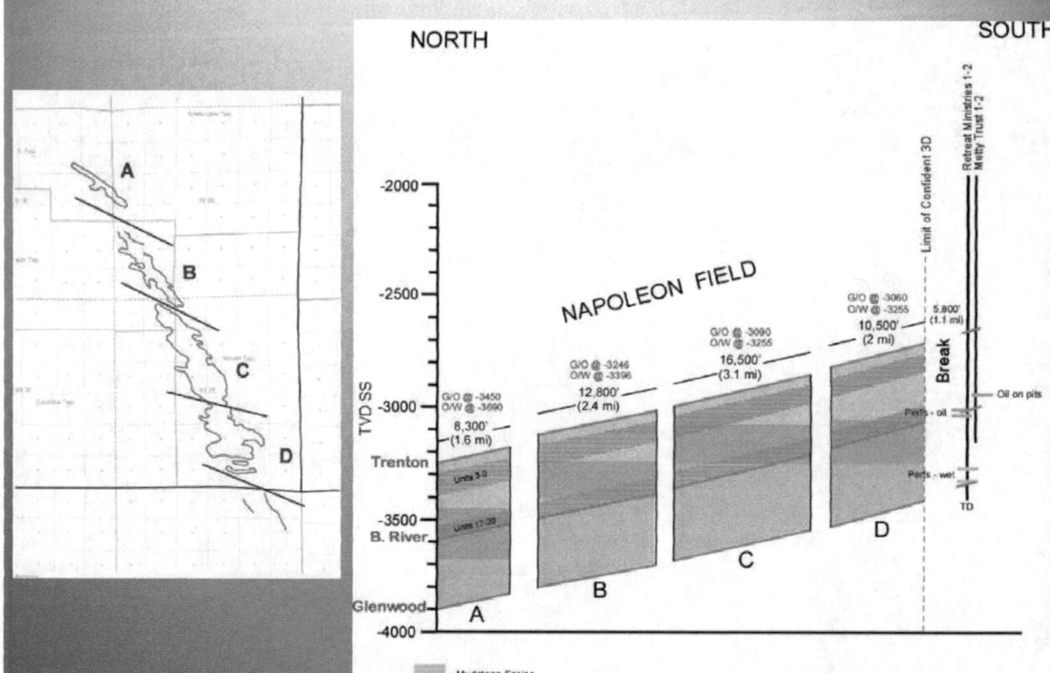

Total 57 wells
53 producers, 4 dry
93% success rate

28 Horizontals, 29 Vertical

Cumulative Production

8,476,602 BO
11,224,259 MCFG
7,306,582 BW

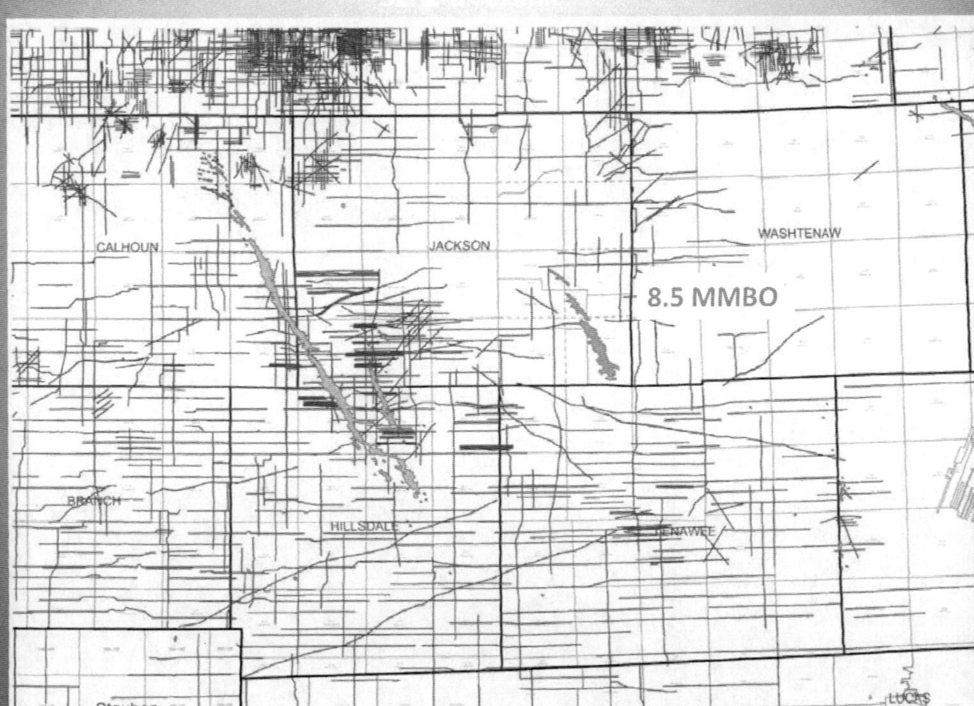

TRENTON-BLACK RIVER
RECENT DISCOVERIES

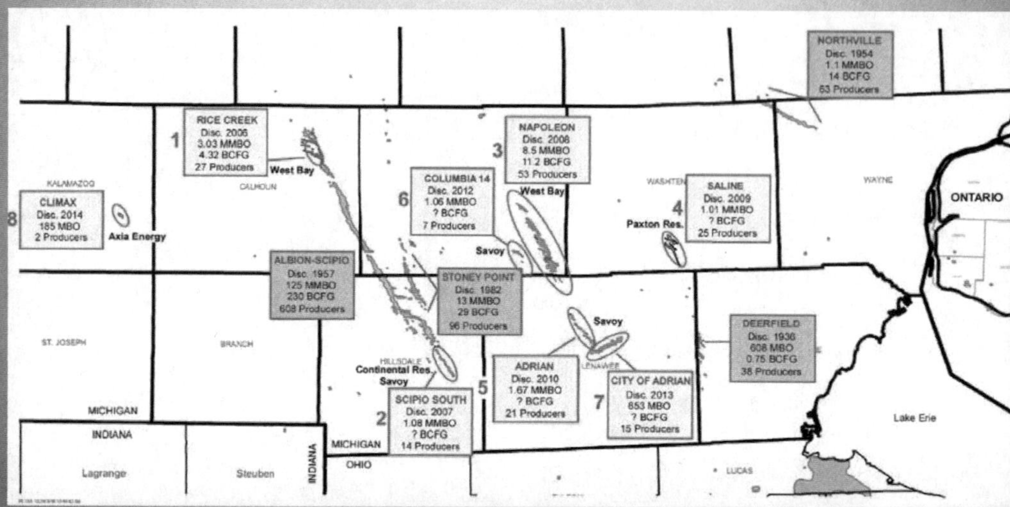

Historic Production = 140 MMBO
Recent Production = 17 MMBO (2D/3D Seismic)

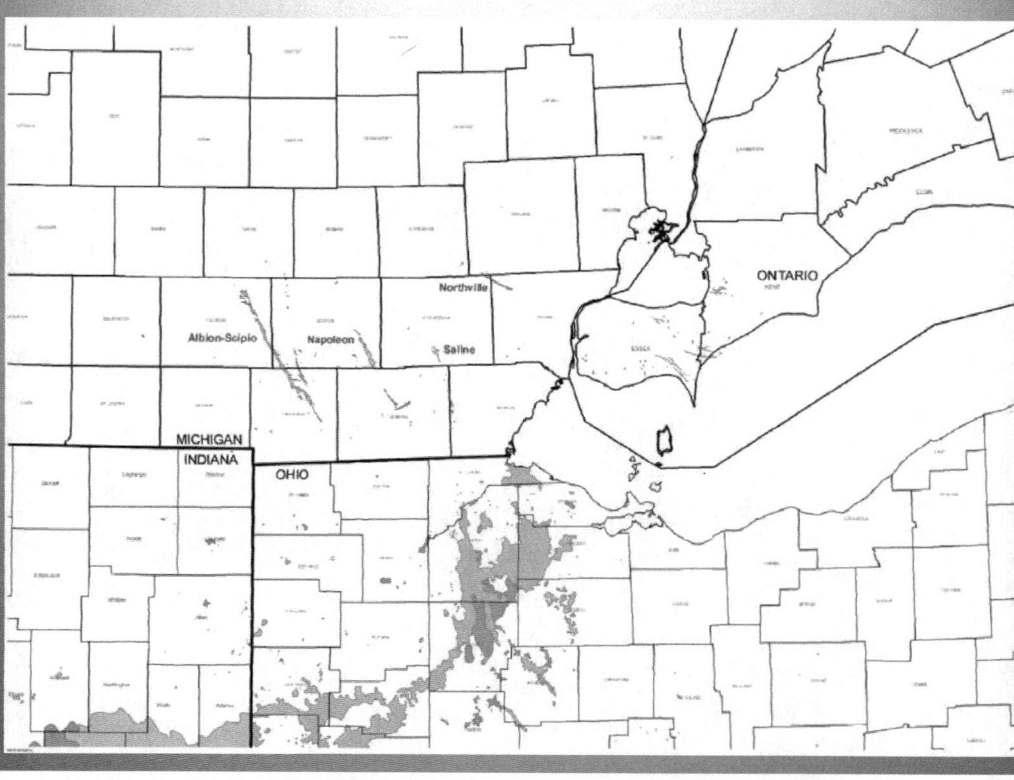

CONFIDENTIAL

Magnetic Survey – Second Vertical Derivative

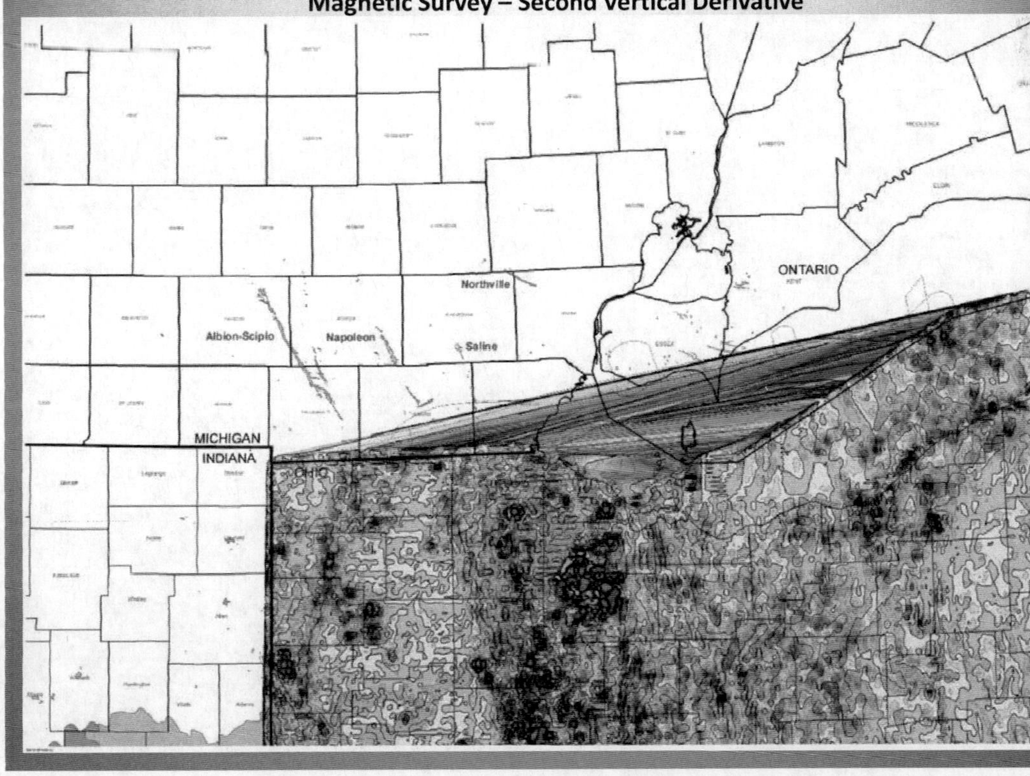

CONFIDENTIAL

CONFIDENTIAL

CONFIDENTIAL

CONFIDENTIAL

CONFIDENTIAL

Devonian Fault Re-activation Scenario

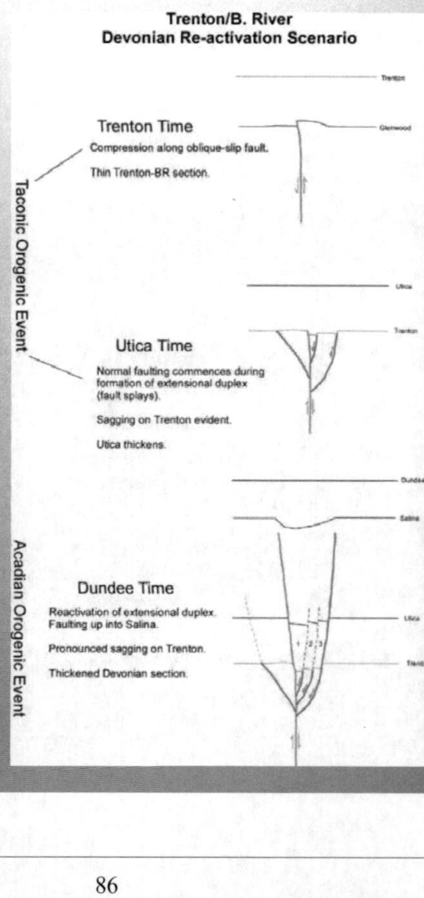

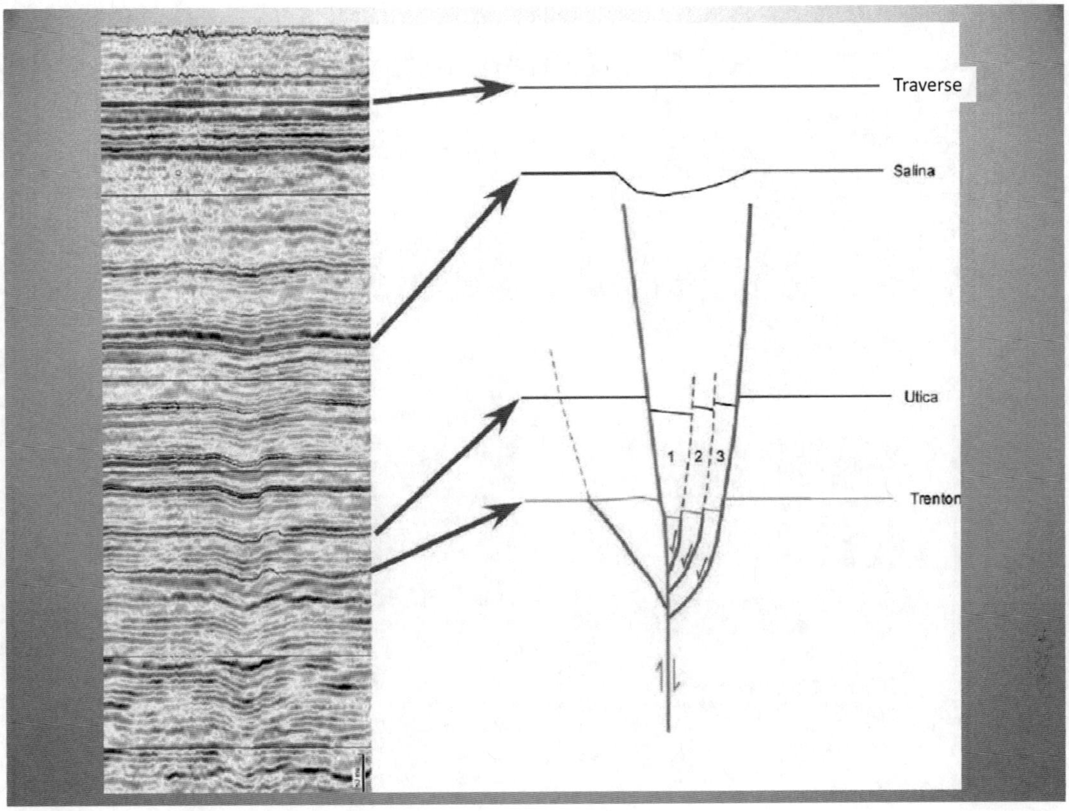

Traverse
Salina
Utica
1
2
3
Trenton

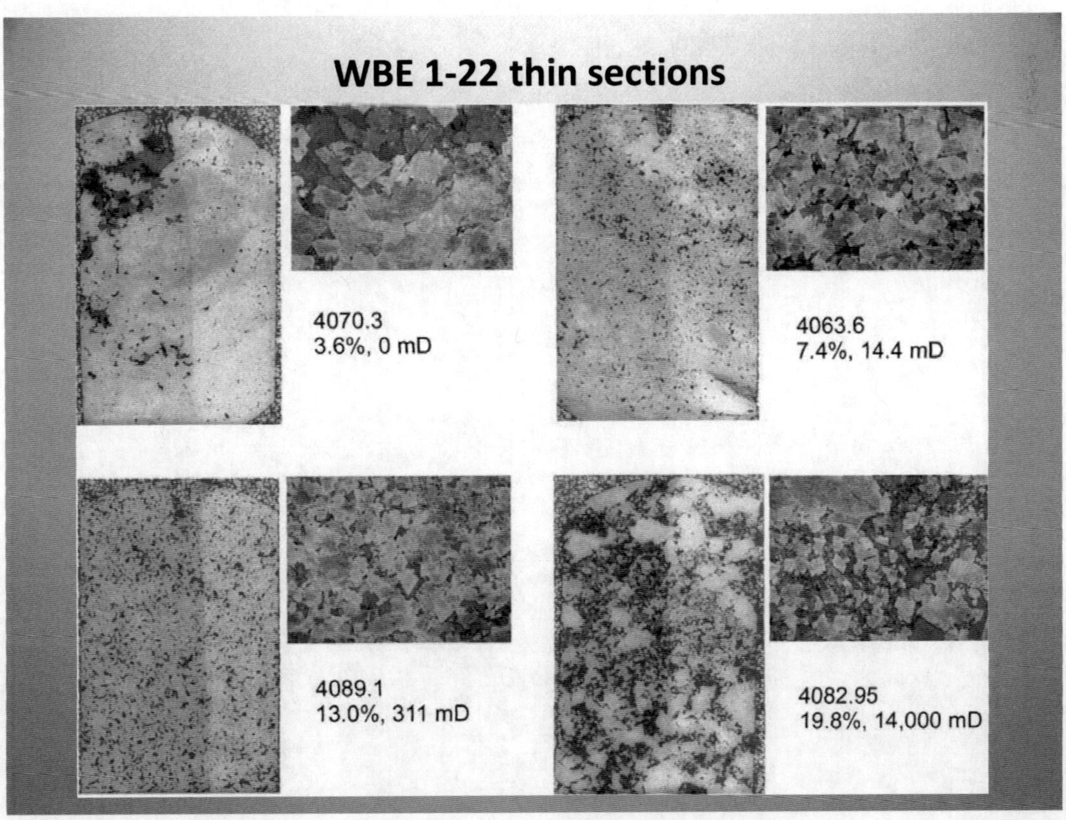

WBE 1-22 thin sections
4070.3
3.6%, 0 mD
4063.6
7.4%, 14.4 mD
4089.1
13.0%, 311 mD
4082.95
19.8%, 14,000 mD

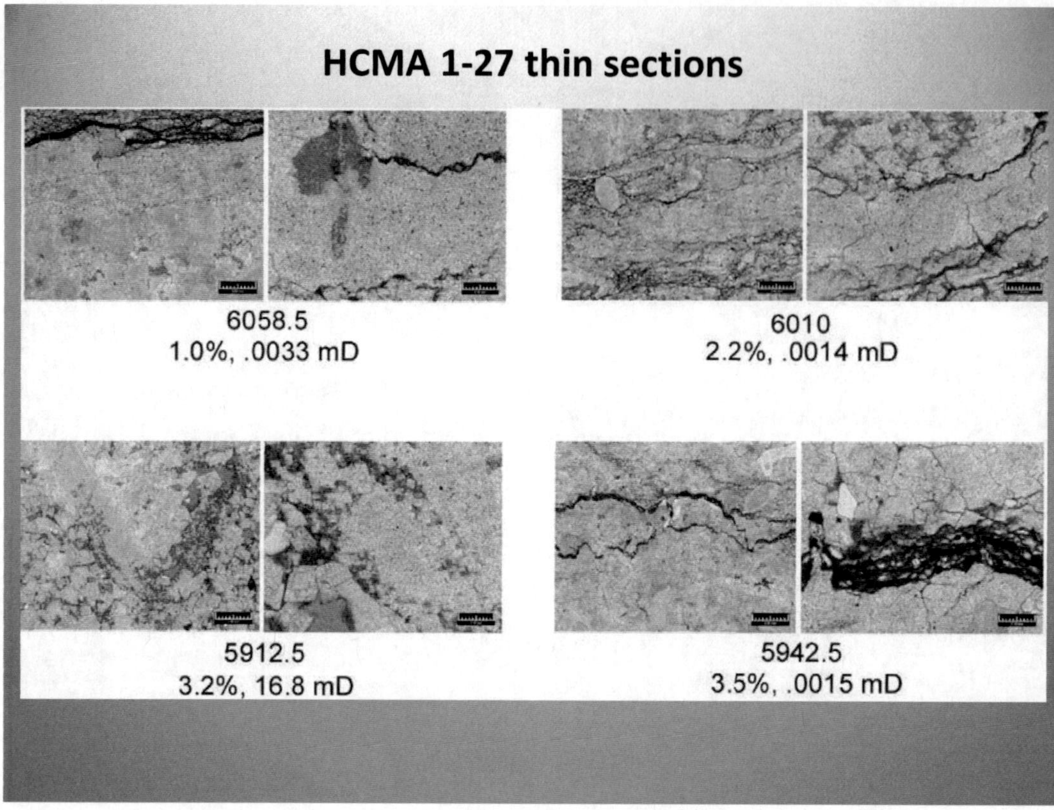

HCMA 1-27 thin sections
6058.5
1.0%, .0033 mD
6010
2.2%, .0014 mD
5912.5
3.2%, 16.8 mD
5942.5
3.5%, .0015 mD

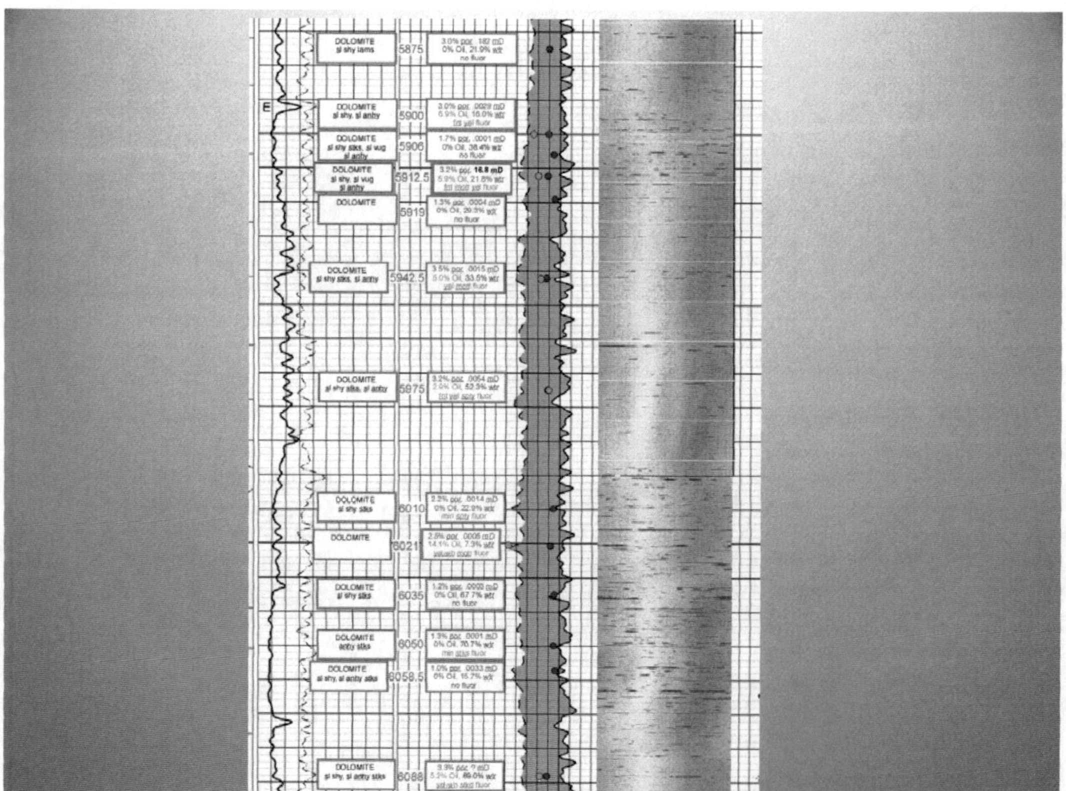

DOLOMITE
sl shy lams
5875
0% Oil, 21.9% wtr
no fluor
DOLOMITE
sl shy, sl anhy
5900
DOLOMITE
5906
1.7% por, .0001 mD
DOLOMITE
sl shy, sl vug
sl anhy
5912.5
3.2% por, 16.8 mD
DOLOMITE
5919
no fluor
DOLOMITE
5942.5
DOLOMITE
5975
DOLOMITE
6010
DOLOMITE
6021
DOLOMITE
6035
no fluor
DOLOMITE
6050
DOLOMITE
6058.5
no fluor
DOLOMITE
6088

HCMA 1-27

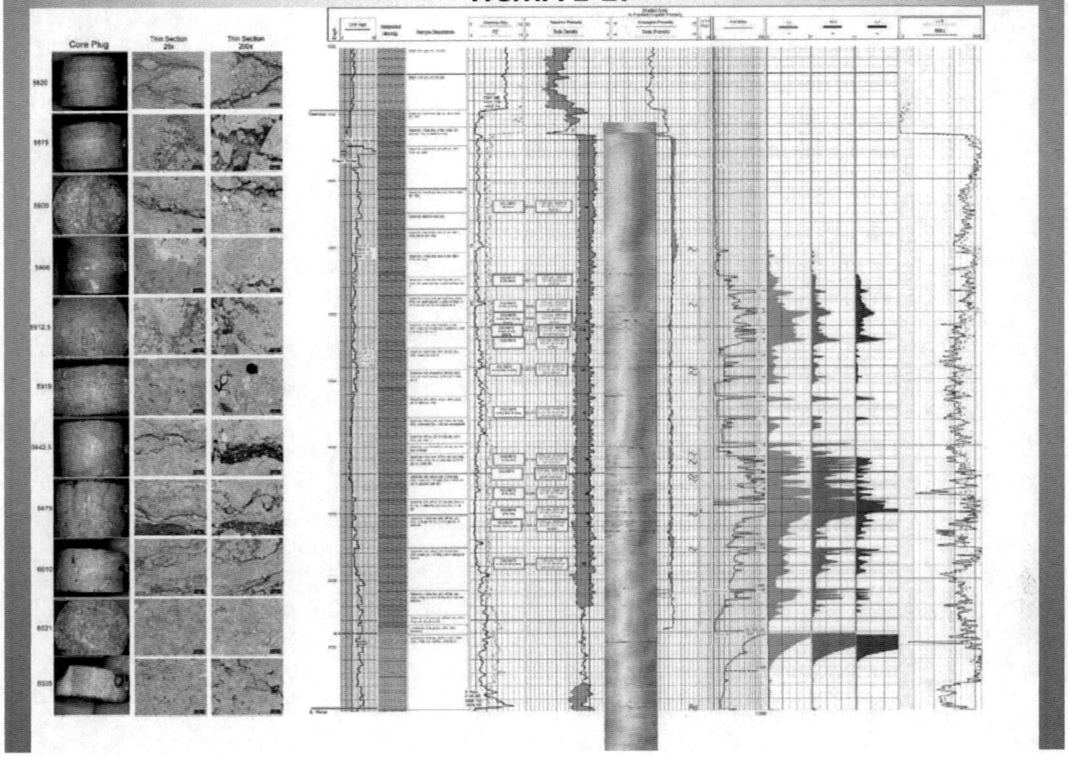

Average porosity/permeabilty in Trenton (core data)

HCMA 1-27 (OAKLAND CO.)

2.3%, 1.3 mD

WBE 1-22 (NAPOLEON FIELD)

6.9%, 901 mD

CONFIDENTIAL

CONCLUSION

- Trenton-Black River discovered in 1936 (Deerfield).

- Albion-Scipio discovered in 1957. It continues to be the largest oilfield in Michigan with cumulative production of 125 MMBO.

- Stoney Point (1985) was the last significant discovery until Rice Creek (2006).

- Rice Creek kicked off a resurgence in Trenton-Black River exploration utilizing both 2D & 3D seismic. Since Rice Creek, seven additional fields have been found cumulating 17 MMBO to date.

- Numerous tight reservoirs have hindered exploration in some areas. This is attributed to occlusion of existing porosity by HTD initiated by Devonian fault re-activation. **TIMING OF FAULTING CAN BE CRUCIAL!**

- Many areas unexplored due to high population density (e.g. Metro Detroit)

West Bay Exploration company

SILURIAN REEF PLAY IN ONTARIO

Allan Phillips (Clinton-Medina Group)

PLAY BY PLAY

Exploring the Michigan Basin

1

MIDDLE AND UPPER SILURIAN CARBONATES AND EVAPORATES

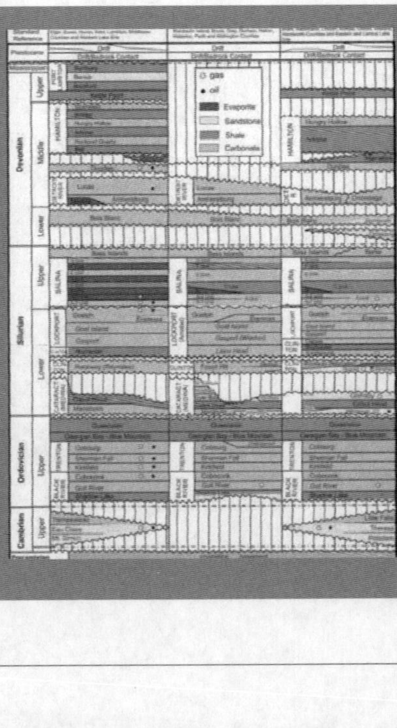

Middle and Upper Silurian age carbonate and evaporite sequence in Lambton, Kent, Essex Counties onshore Ontario and offshore in the western and central basins of Lake Erie.

Guelph-Lockport Group reef buildups sealed with evaporites of the overlying Salina Group.

Structural and stratigraphic traps of banded carbonates within the basal portion of the Salina Group (A1 Carbonate and A2 Carbonate)

2

FOCUS ON THE SOUTHEASTERN RIM OF MICHIGAN BASIN

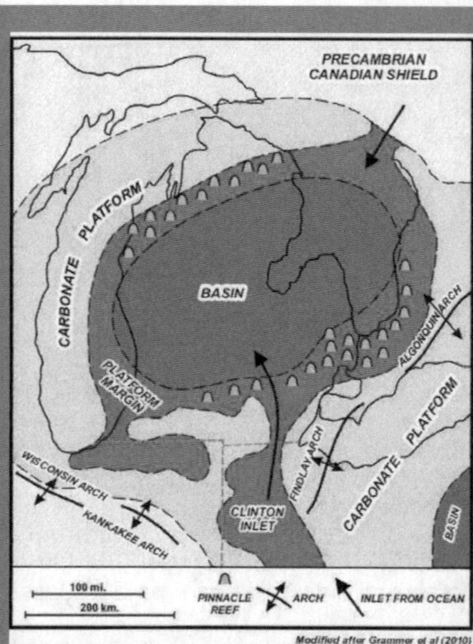

- Silurian carbonate reservoirs on the southeastern rim of the Michigan Basin in southwestern Ontario.
- Shallow drill depth to top of reservoir is 230-780m (750-2560').
- Natural gas is the predominate hydrocarbon trapped in these reservoirs that straddle basin rimming arches.

3

THE EARLY DAYS

Eugene Coste "Father of Natural Gas in Canada"

4

EARLY GUELPH OIL AND GAS DISCOVERIES

1.Coste No.1, (Kingsville Pool), January 23, 1889 struck gas at 1020' flowing at 10 MMcf/d.

2. Oil discovered on Pelee Island in 1895. Most wells were modest producers 2-6 bopd with some gas at a depth of 740-750'.

3. Acme Oil No.1 (Fletcher Pool), December , 1905 struck gas at 1360' and oil @ 1385' flowed oil at 40 bopd.

4. Fairbank Gusher (Oil Springs), March 7, 1914, struck gas at 1900' flowing 11 MMcf/d.

5

GUELPH (NIAGARAN) REEF TRENDS

Barrier-Patch Reef Complex

Pinnacle Reef Belt

Pinnacle Reef > 50m (164')

Incipient Reef < 50m (164')

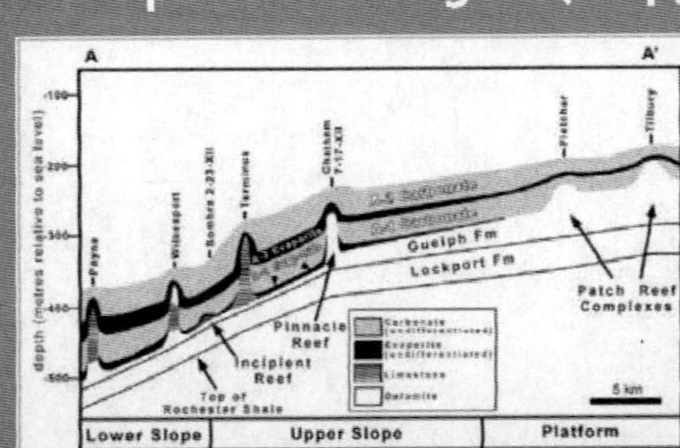

6

GUELPH PATCH-BARRIER REEFS

Buildups on the basin rimming carbonate complex.

Kingsville-Leamington, Fletcher, Tilbury, D'Clute (1889-1910)

wildcat drilling, well witching, divining, trendology

Tilbury-D'Clute Offshore, Glasgow-Talbot (1913-1977)

trendology, shallow structure mapping, gravity, seismic

7

GUELPH PATCH-BARRIER REEFS WEST & CENTRAL LAKE ERIE

Natural gas is trapped in the crest of reef buildups along the 100 plus mile platform reef complex.

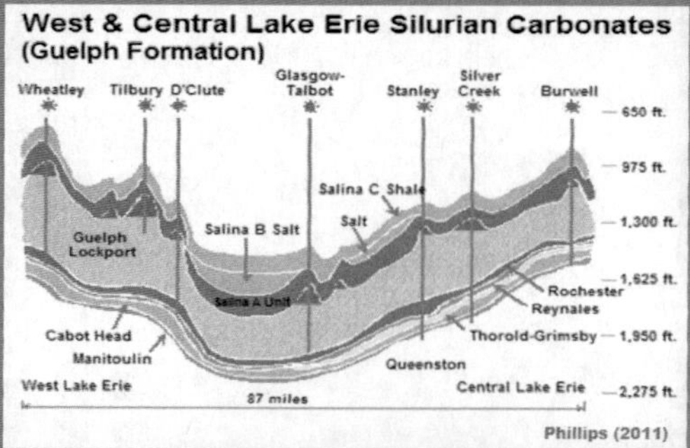

8

GLASGOW-TALBOT GAS POOL

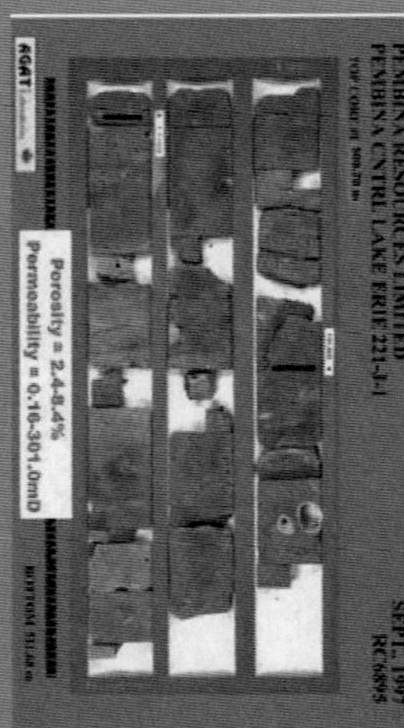

9

GLASGOW-TALBOT GAS POOL, CENTRAL LAKE ERIE

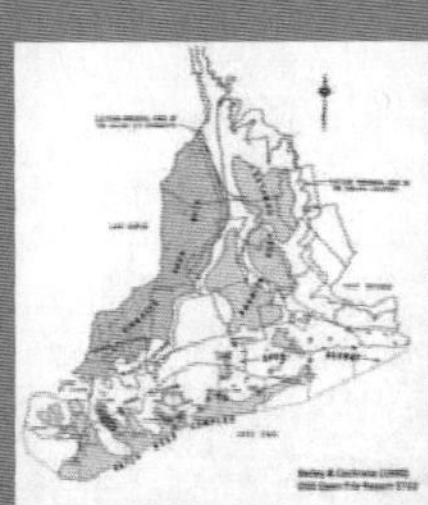

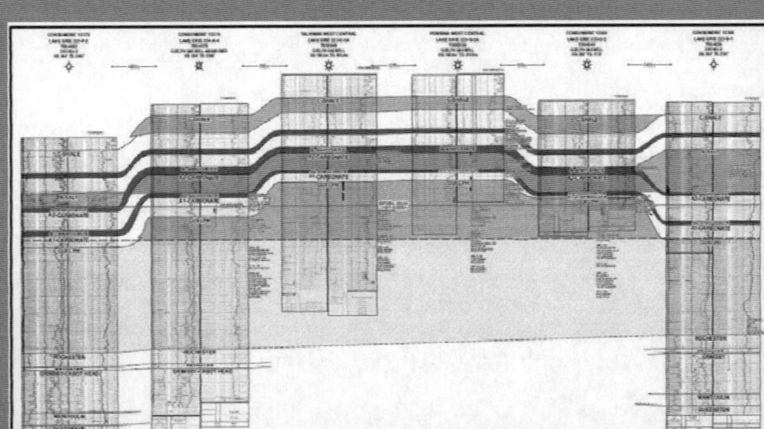

Total Guelph A1 Carbonate build up > 170m (558′).

Regional thickness = 100m (328′)

10

GLASGOW-TALBOT GAS POOL UPSIDE UPHOLE IN A2 CARBONATE

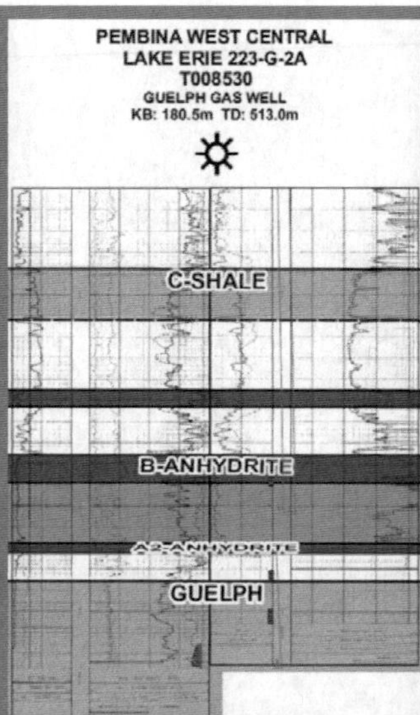

Infill gas well drilled in 1997. Gas show @ 469m (1539') in Salina A2 Carbonate section.

DST #2 flowed gas at 4,885 m3/d (172 Mcf/d), SIP = 4502 kPa (653 psig)

Guelph reef section below is the better reservoir ,testing gas at 44,098 m3/d (1557 Mcf/d), was partially depleted SIP = 3438 kPa (499 psig)

11

HORIZONTAL DRILLING A2 CARBONATE RESERVOIR

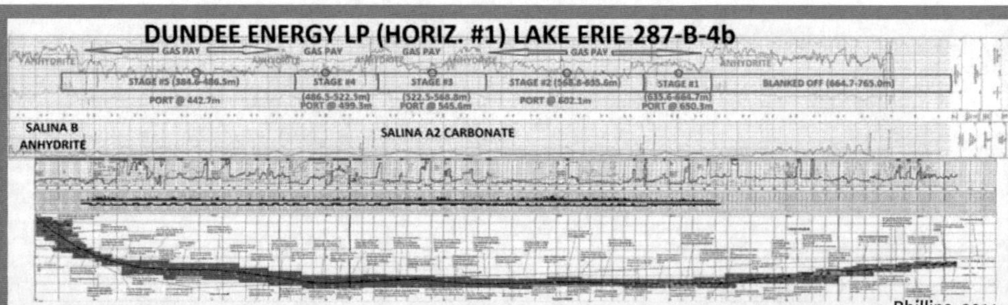

Phillips, 2014

Offshore Tilbury Pool, Lake Erie 287-B-4b

Drilled 328m (1076') of horizontal section

5 stages acidized with 7-20 m3 (45-60 bbls)

Flow to cleanup at 8,496 m3/d (300 Mcf/d)

12

GUELPH PINNACLE AND INCIPIENT REEFS

Buildups moving deeper in the Michigan Basin.

Oil Springs incipient (1914), Dawn 47-49 pinnacle (1930)

wildcat drilling, well witching, divining, shallow structure mapping

Kimball-Colinville, Payne, Corunna, Moore 3-21-XII (1947-1977)

shallow structure mapping ,gravity, seismic

13

GUELPH PINNACLE-INCIPIENT REEFS LAMBTON & HURON COUNTIES

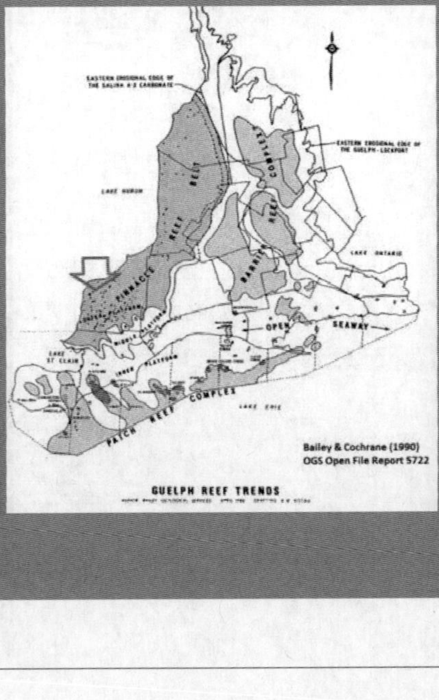

Pinnacle Reef Belt is basinward from the Patch-Barrier Complex that rims the Michigan Basin.

Oil and gas pools in these discrete reefs can be prolific, but the upside is in the reservoir can be reused as a gas storage pool.

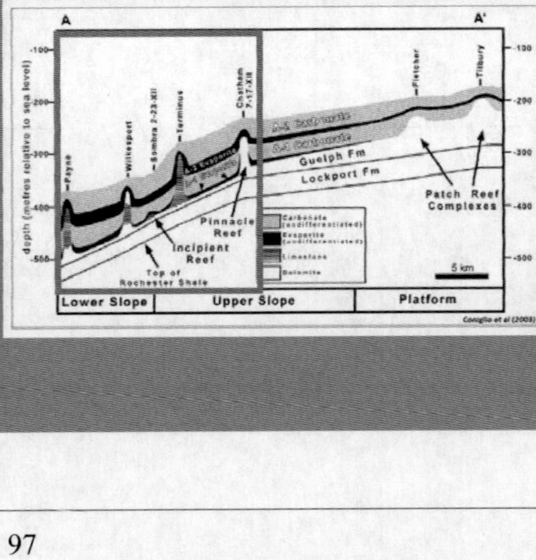

14

RATE OF DISCOVERY OF SILURIAN REEFS IN ONTARIO

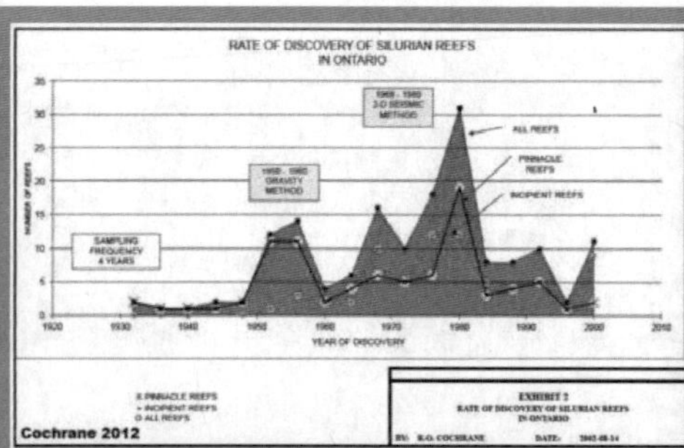

1950-1960: Gravity Method, looking for salt thinning

1968-1980: 2D Seismic Method, looking for reef buildups

15

PINNACLE & INCIPIENT REEF WAUBANO & DOW MOORE 7-6-III

Waubano Pinnacle Reef 450′ (137.2m), 9.85 Bcf Gas

Moore 7-6-III Incipient Reef 120′ (36.6m), 55,835 bbls Oil

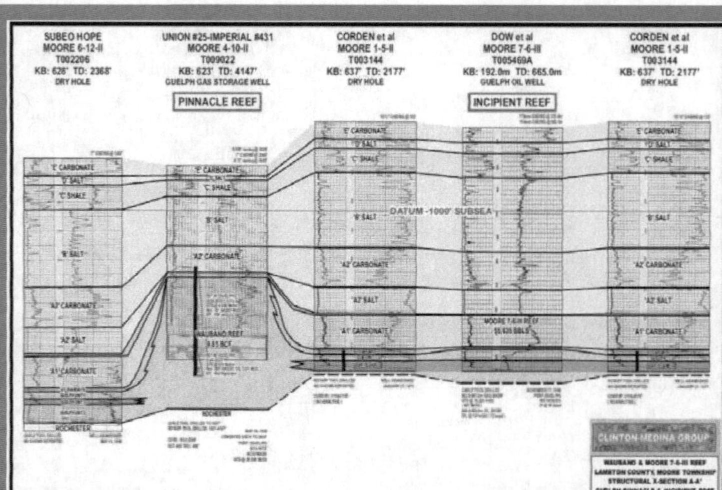

16

GAS STORAGE RESERVOIR MOORE 3-21-XII

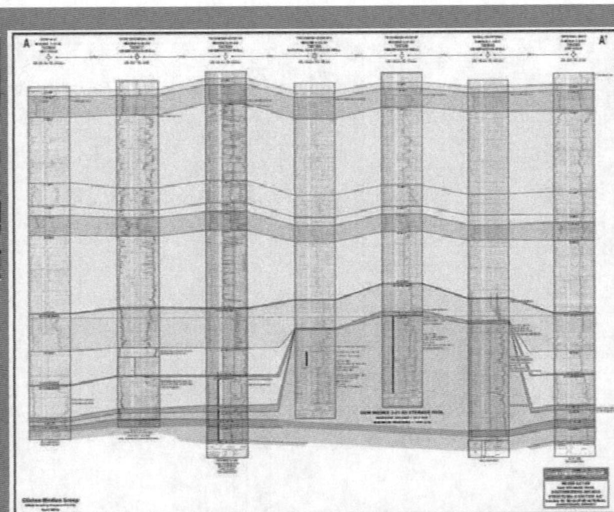

Discovery well Dow Moore 3-21-XII drilled in June 1977, flowed GTS @ 31 MMcf/d (877,920 m3/d) from a depth of 2471' (753.2m)

17

GAS STORAGE RESERVOIR ATTRIBUTES

Excellent limestone reservoir sealed with a cap of impermeable anhydrite

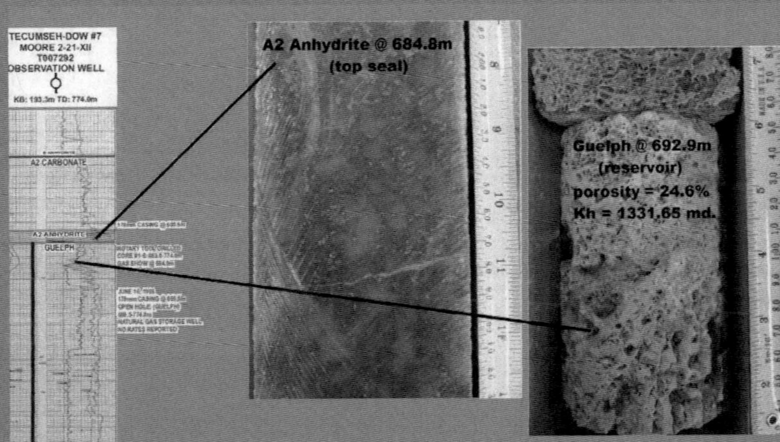

18

CORUNNA GAS STORAGE POOL OIL LEG EOR WITH GAS INJECTION

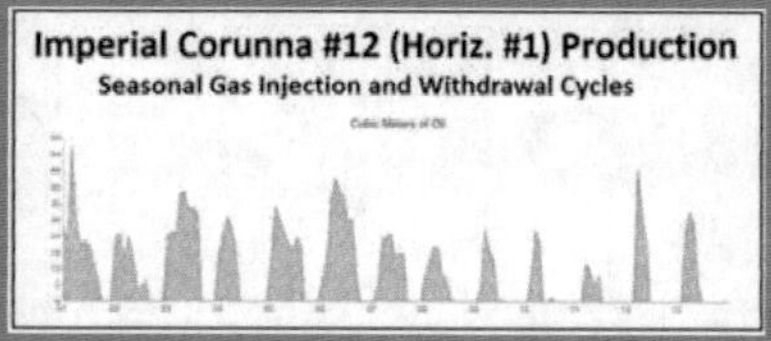

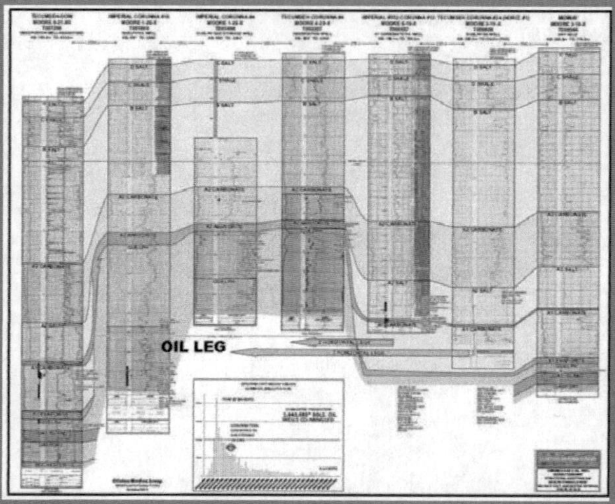

19

GUELPH AND SALINA STRUCTURAL POOLS

Oil & gas pools along fault trends.

Dawn , Chatham-Dresden-Camden Gore, Zone (1914-1943)

wildcat drilling, well witching, divining, trendology

Becher East, Becher West (1946)

shallow structure mapping , trendology

20

GUELPH-SALINA STRUCTURAL OIL AND GAS TRAPS (BECHER WEST)

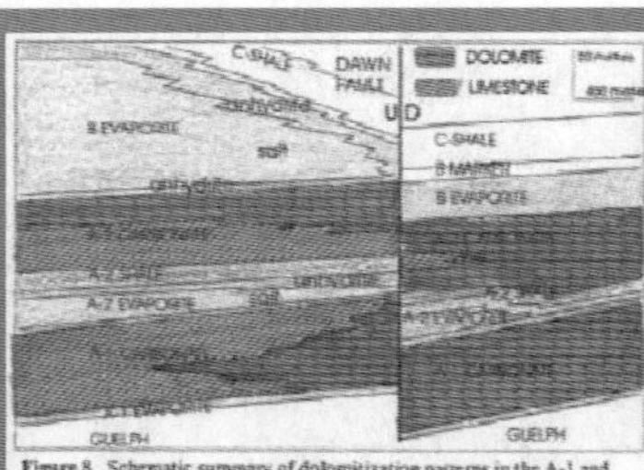

Figure 8. Schematic summary of dolomitization patterns in the A-1 and A-2 Carbonate Units near the Dawn Fault in Sombra Township, Lambton County, in southwestern Ontario. Modified from Carter (1991)

Carter et al (1994)

Discovery well in Becher West Pool, Imperial #15 Sombra 8-9-V completed in February 1946 produced oil at 15 bopd from the Salina A1 Carbonate at 1844' (562.1m)

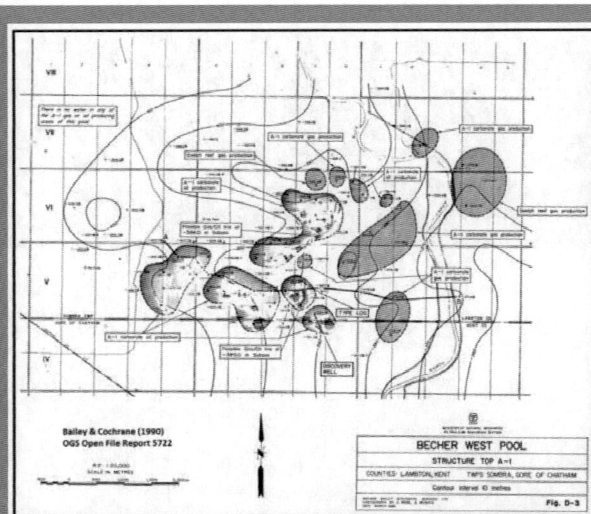

21

BECHER WEST OIL POOL 2,679,842 BBLS RECOVERABLE

Very good porosity with fair permeability in this very fine grained dolomite with 36.0°API oil was converted to water flood in 1964

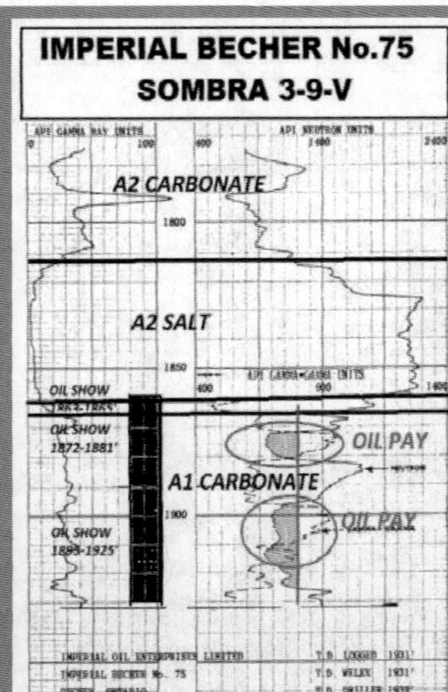

IMPERIAL BECHER #75
SOMBRA 3-9-V
UPPER POROUS RESERVOIR
OIL STAINED DOLOMITE
POROSITY = 18.7%
PERMEABILITY = 33.0md

22

WHERE DO WE GO FROM HERE?

Looking at the historical production can we predict where to go looking for more production from these reservoirs?

23

TABULATION OF NATURAL GAS TRAPS IN SILURIAN CARBONATES

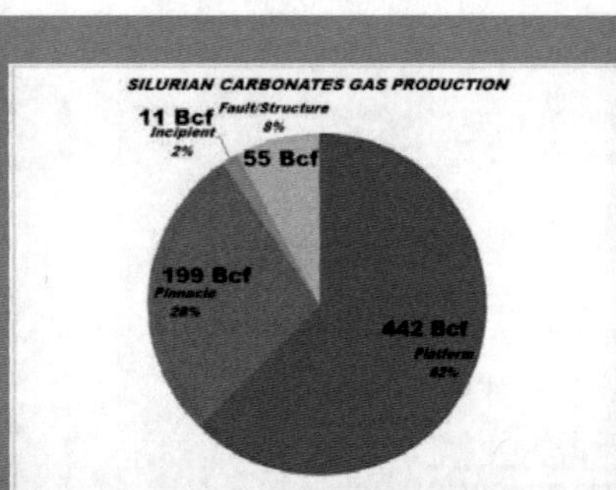

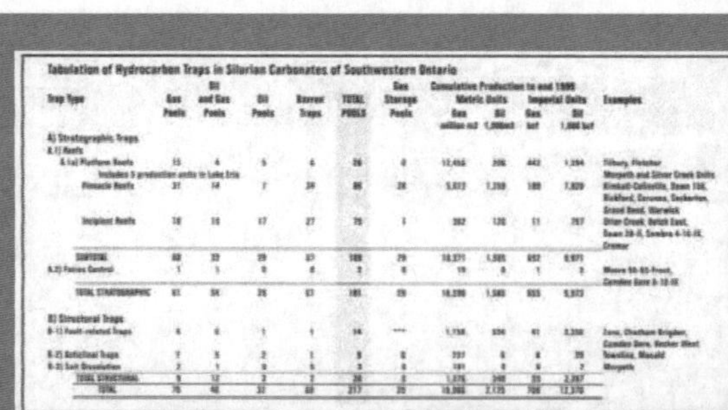
Tabulation of Hydrocarbon Traps in Silurian Carbonates of Southwestern Ontario

Cochrane, 2004

708 Bcf (19,966 million m3) of gas production to Q4/1999.
Majority of the gas production (62%) is from platform reefs on the up dip margin of the basin. Moving basinward, pinnacle reefs account for next highest percentage (28%) and incipient reefs for 2%. Structural Guelph/Salina traps make up the balance (8%).

24

TABULATION OF OIL TRAPS IN SILURIAN CARBONATES

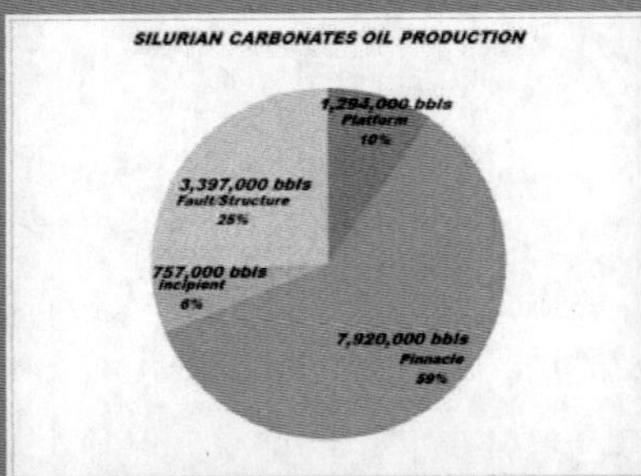

Tabulation of Hydrocarbon Traps in Silurian Carbonates of Southwestern Ontario

Trap Type	Gas Pools	Oil and Gas Pools	Oil Pools	Barren Traps	TOTAL POOLS	Gas Storage Pools	Cumulative Production to end 1999 Metric Units Gas million m3	Oil 1,000m3	Imperial Units Gas bcf	Oil 1,000 bbl	Examples
A) Stratigraphic Traps											
A.1) Reefs											
A.1a) Platform Reefs	13	4	5	6	28	4	12,466	206	442	1,294	Tilbury, Fletcher
Includes 5 production units in Lake Erie											
Pinnacle Reefs	31	18	7	34	88	24	5,613	1,259	199	7,920	[illegible] and Silver Creek Units, Kimball-Colinville, Dawn 156, Bickford, Corunna, Seckerton, Grand Bend, Warwick
Incipient Reefs	18	11	17	27	73	1	302	120	11	757	[illegible]
SUBTOTAL	62	33	29	67	189	29	18,371	1,585	652	9,971	
A.2) Facies Control	1	1	0	0	2	0	19	0	1	2	[illegible]
TOTAL STRATIGRAPHIC	63	34	29	67	191	29	18,390	1,585	653	9,973	
B) Structural Traps											
B-1) Fault-related Traps	6	6	1	1	14	—	1,156	534	41	3,358	[illegible]
B-2) Anticlinal Traps	1	5	2	1	9	0	251	6	9	39	[illegible]
B-3) Salt Dissolution	2	1	0	0	3	0	167	0	6	2	[illegible]
TOTAL STRUCTURAL	9	12	3	2	26	0	1,574	540	55	3,397	
TOTAL	72	46	32	69	217	29	19,964	2,125	708	13,370	

Cochrane, 2004

13,370,000 bbls (2,125,000 m3) of oil production to Q4/1999.
Majority of the oil production (59%) is from pinnacle reefs deeper in the basin. Structural Guelph/Salina traps account for next highest percentage (25%) and incipient reefs for 6%. Platform reefs (Fletcher Pool) make up the balance (10%).

25

OIL & GAS IN SILURIAN CARBONATES OF THE GUELPH AND LOWER SALINA

- Historical production from these shallow Silurian carbonate pools in southwestern Ontario on the southeastern rim of the Michigan Basin have had a long and storied history.
- Some of these prolific reservoirs are being reused as natural gas storage pools in the second largest storage hub in North America.
- Exploration for smaller lower productivity gas reservoirs is faced with challenging economics with low gas prices and higher risk.
- Economics for oil is more favorable, therefore focus should be on exploration and development of onshore oil prospects.
- Exploration for Salina and Guelph structures, incipient reefs and enhanced oil recovery in existing Guelph reef and structural oil pools should be the focus in this current economic environment.

26

REFERENCES

- Bailey Geological Services Ltd. and Cochrane, R.O. 1990, Geology of selected oil and gas pools in the Silurian carbonates of southern Ontario; Ontario Geological Survey, Open File Report 5722, 50p.
- Cochrane, R.O. 2004, Reservoirs and Production from the Silurian Carbonate Rocks. Ontario Petroleum Institute Magazine, Play-by-Play Action! Ontario Oil & Gas 2004. Ontario Petroleum Institute, London, Ontario.
- Cochrane, R.O. 2012, Reservoirs and Production from the Silurian Carbonate Rocks of Southwestern Ontario. Ontario Petroleum Institute Magazine, 50th Anniversary Edition Ontario Oil & Gas 2012, DEL Communications Inc. Winnipeg, Manitoba.
- Coniglio, M., Zheng, Q., and Carter, T.R., 2003, Dolomite and recrystallization of middle Silurian reefs and platformal carbonates of the Guelph Formation, Michigan Basin, southwestern Ontario. Bulletin of Canadian Petroleum Geology, v.51, p. 177-199.
- Carter, T.R., Trevail, R. A., and Smith, L. 1994. Core workshop: Niagaran Reef and inter-reef relationships in the subsurface of southwestern Ontario; Geological Association of Canada 1994 Annual Meeting, Waterloo, Ontario, Field Trip A5 Guidebook, 37 p.
- Grammer, M.G., Noack, A.K., Qualman, H., Ritter-Varga, A., Wold, J., Sandomierski, A.E., and Harrison III, W.B., 2010, Reservoir Characterization of Silurian (Niagaran) "Pinnacle" Reefs in the Michigan Basin. Oral Presentation at the AAPG Annual Convention and Exhibition, New Orleans, Louisiana, USA, Search and Discovery Article #50286.
- Newton, A. C. 1958, Offshore Exploration for Gas Under the Canadian Waters of the Great Lakes. Ontario Department of Mines, Geological Circular No. 7, 37p.
- Phillips, A. R., 2011, Drilling in Lake Erie, Pushing 100 Years. Oral Presentation at the Ontario Petroleum Institute 50th Annual Conference and Trade Show, London, Ontario, Canada, October 19-21, 2011
- Phillips, A. R., 2014,100 Years of Offshore Drilling, The Changing Technology Used in Lake Erie Drilling. Oral Presentation at the AAPG Eastern Section Meeting, London, Ontario, Canada, September 27-30, 2014.

2017 OPI Conference

Economics of Various Play Types in SW Ontario

By: Jim McIntosh

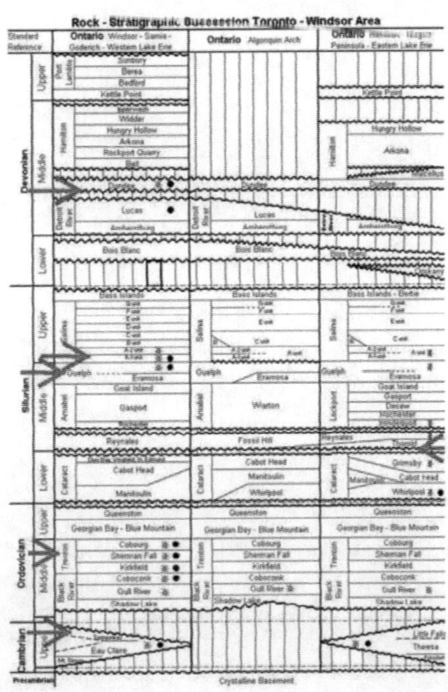

"Thumb-Nail Economic Models"

- Devonian reservoir: Oil with water flooding, vertical wells only and vertical/horizontal wells
- Silurian A-1 Carb reservoirs: Gas, vertical well versus horizontal well development
- Silurian Guelph reservoirs: Gas partially salt-plugged pinnacle reef, Oil partially salt-plugged pinnacle reef
- Silurian Clinton reservoirs: Gas, vertical well versus horizontal well development
- Ordovician T/BR reservoirs: Gas and Oil, vertical well versus horizontal well development
- Cambrian reservoirs: Gas and Oil, vertical well versus horizontal well development

Reserve Calculation Table

Geological Formation	Assumed Depth		Ave Ø	Water Sat'n	Res Press		Res Temp		Oil FVF	Gas Compr	OOIP Factor		OGIP Factor	
	(m)	(ft)	(%)	(%)	(kPaa)	(psia)	(°C)	(°F)	(Rm^3/STm^3)		(m3/ha-m)	(Bbl/ac-ft)	(103m3/ha-m)	(Mcf/ac-ft)
Devonian	122	400	19%	6%	1,269	184	17	63	1.05		1,713	1,329		
Silurian A-1 Gas	425	1394	14%	15%	4,250	616	18	64.4		0.87			56.8	249
Silurian Guelph Oil	600	1969	8%	25%	6,000	870	19	66.2	1.05		571	443		
Silurian Guelph Gas	575	1887	8%	15%	5,750	834	14	57.2		0.85			45.6	199
Silurian Clinton Gas	420	1378	12%	10%	4,200	609	16	60.8		0.88			50.7	222
Ordovician Gas	900	2953	10%	10%	9,000	1,305	18	64.4		0.81			97.7	428
Ordovician Oil	830	2723	10%	10%	8,300	1,204	18	64.4	1.15		783	607		
Cambrian Gas	1189	3900	8%	20%	11,887	1,724	21	69.8		0.78			94.3	413
Cambrian Oil	1189	3900	8%	20%	11,887	1,724	21	69.8	1.2		533	414		

Devonian pool development

D&A costs: Vertical Devonian-depth well	$80k/well
D&A costs: Horizontal Devonian-depth well	$300k/well
Completion costs: Vertical Devonian-depth well	$20k/well
Completion costs: Horizontal Devonian-depth well	$50k/well
Equipping costs: Vertical Devonian-depth Producing well	$10k/well
Equipping costs: Horizotal Devonian-depth Producing well	$30k/well
Equipping costs: Injection well	$8k/well
Pipelines: Average/well costs	$10k/well
Battery Costs (Disp well, water source well, injection pump, tanks):	$500k
Abandonment costs: vertical well	$3k/well
Abandonment costs: horizontal well	$30k/well
Abandonment costs: battery	$50k

Silurian A-1/Guelph/Clinton pool development

D&A costs: Vertical Silurian/Guelph/Clinton-depth well	$200k/well
D&A costs: Horizontal Silurian/Guelph/Clinton-depth well	$400k/well
Completion costs: Vertical Silurian/Guelph-depth well	$50k/well
Completion costs: Vertical Clinton-depth well	$100k/well
Completion costs: Horizontal Silurian/Guelph-depth well	$150k/well
Completion costs: Horizontal Clinton-depth well	$200k/well
Equipping costs: Vertical Silurian/Guelph/Clinton-depth Producing Gas well	$10k/well
Equipping costs: Vertical Silurian/Guelph-depth Producing Oil well	$30k/well
Equipping costs: Horizotal Silurian/Guelph/Clinton-depth Producing Gas well	$30k/well
Equipping costs: Horizotal Silurian/Guelph-depth Producing Oil well	$50k/well
Equipping costs: Injection well	$8k/well
Pipelines: Average/well costs	$10k/well
Battery Costs (Disp well, water source well, injection pump, tanks):	$500k
Compressor site costs (Disp well, compressor, TEG dehy, meter site):	$400k
Abandonment costs: vertical gas well	$10k/well
Abandonment costs: vertical oil well	$60k/well
Abandonment costs: horizontal gas well	$20k/well
Abandonment costs: horizontal oil well	$150k/well
Abandonment costs: battery	$150k
Abandonment costs: Compressor Site	$75k

Ordovician pool development

D&A costs: Vertical Ordovician-depth well	$400k/well
D&A costs: Horizontal Ordovician-depth well	$900k/well
Completion costs: Vertical Ordovician-depth well	$100k/well
Completion costs: Horizontal Ordovician-depth well	$200k/well
Equipping costs: Vertical Ordovocian-depth Producing Gas well	$30k/well
Equipping costs: Vertical Ordovician-depth Producing Oil well	$75k/well
Equipping costs: Horizotal Ordovician-depth Producing Gas well	$40k/well
Equipping costs: Horizotal Ordovician-depth Producing Oil well	$100k/well
Pipelines: Average/well costs	$100k/well
Battery Costs (Disp well, water source well, injection pump, tanks):	$500k
Compressor site costs (Disp well, compressor, TEG dehy, meter site):	$400k
Abandonment costs: vertical gas well	$60k/well
Abandonment costs: vertical oil well	$60k/well
Abandonment costs: horizontal gas well	$150k/well
Abandonment costs: horizontal oil well	$150k/well
Abandonment costs: battery	$150k
Abandonment costs: Compressor Site	$75k

Cambrian pool development

D&A costs: Vertical Cambrian-depth well	$400k/well
D&A costs: Horizontal Cambrian-depth well	$900k/well
Completion costs: Vertical Cambrian-depth well	$100k/well
Completion costs: Horizontal Cambrian-depth well	$200k/well
Equipping costs: Vertical Cambrian-depth Producing Gas well	$30k/well
Equipping costs: Vertical Cambrian-depth Producing Oil well	$75k/well
Equipping costs: Horizotal Cambrian-depth Producing Gas well	$40k/well
Equipping costs: Horizotal Cambrian-depth Producing Oil well	$100k/well
Equipping costs: Cambrian-depth Injection well	$75k/well
Pipelines: Average/well costs	$100k/well
Battery Costs (Disp well, water well, injection pump, tanks, compressor):	$700k
Compressor site costs (Disp well, compressor, TEG dehy, meter site):	$400k
Abandonment costs: vertical gas well	$60k/well
Abandonment costs: vertical oil well	$60k/well
Abandonment costs: horizontal gas well	$150k/well
Abandonment costs: horizontal oil well	$150k/well
Abandonment costs: battery	$150k
Abandonment costs: Compressor Site	$75k

Natural Gas Price Forecast

Year	Average NYMEX H Hub ($US/MMBtu)	Average Exchange ($US/$Cdn)	Basis to NYMEX ($Cdn/MMBtu)	Ontario Price ($Cdn/MMBtu)	Ont Gas Price ($Cdn/Mcf)
2011 Act	$4.02	$1.012	-$0.71	$3.26	$3.42
2012 Act	$2.79	$1.001	$0.35	$3.14	$3.29
2013 Act	$3.72	$0.971	-$0.09	$3.74	$3.92
2014 Act	$4.30	$0.906	$0.52	$5.27	$5.52
2015 Act	$2.62	$0.782	$0.08	$3.43	$3.59
2016 Act	$2.52	$0.757	$0.05	$3.37	$3.53
2017	$3.50	$0.757	$0.03	$4.65	$4.87
2018	$3.35	$0.786	$0.03	$4.29	$4.50
2019	$3.50	$0.813	$0.03	$4.34	$4.55
2020	$3.79	$0.831	$0.03	$4.59	$4.81
2021	$4.00	$0.850	$0.03	$4.73	$4.96
2022	$4.12	$0.850	$0.03	$4.87	$5.11
2023	$4.24	$0.850	$0.03	$5.02	$5.26
2024	$4.35	$0.850	$0.03	$5.15	$5.39
2025	$4.42	$0.850	$0.03	$5.24	$5.49
2026	$4.52	$0.850	$0.03	$5.35	$5.60
2027	$4.61	$0.850	$0.03	$5.46	$5.72
2028	$4.70	$0.850	$0.03	$5.57	$5.83
2029	$4.79	$0.850	$0.03	$5.67	$5.94
2030	$4.89	$0.850	$0.03	$5.78	$6.06
2031	$4.98	$0.850	$0.03	$5.90	$6.18

Crude Oil Price Forecast

Year	Average NYMEX WTI ($US/Bbl)	Average Exchange ($US/$Cdn)	Average Edm Par ($Cdn/Bbl)	Basis WTI-Edm ($US/Bbl)	Basis Srns-Edm ($Cdn/Bbl)	Brk/War/P Est Oil Price ($Cdn/Bbl)	Basis Srns-Edm ($Cdn/Bbl)	Civle/Willey Oil Price ($Cdn/Bbl)	Basis Srns-Edm ($Cdn/Bbl)	Rom 2-215-II Oil Price ($Cdn/Bbl)	Basis Srns-Edm ($Cdn/Bbl)	Rodney Oil Price ($Cdn/Bbl)	Basis Srns-Edm ($Cdn/Bbl)	Dover E Oil Price ($Cdn/Bbl)
2011 Act	$95.07	$1.012	$95.23	-$1.30	-$1.11	$94.12	$1.46	$96.69	$0.80	$96.03	$0.75	$95.98	$0.08	$95.31
2012 Act	$94.20	$1.001	$86.42	$7.73	-$1.16	$85.26	$3.53	$89.95	$2.78	$89.21	$2.62	$89.04	$3.69	$90.11
2013 Act	$97.96	$0.971	$93.26	$7.40	$0.30	$93.57	$6.09	$99.35	$11.67	$104.93	$2.04	$95.30	$5.15	$98.42
2014 Act	$93.00	$0.906	$94.02	$7.85	$1.31	$95.34	$6.16	$100.19	$9.07	$103.10	-$0.66	$93.36	$2.46	$96.48
2015 Act	$48.79	$0.782	$57.47	$3.84	-$0.32	$57.15	$5.98	$63.44	$5.17	$62.64	$1.22	$58.68	$6.00	$63.47
2016 Act	$43.31	$0.757	$53.18	$3.05	-$0.45	$52.73	$6.70	$59.89	$5.86	$59.04	$1.64	$54.82	$5.78	$58.97
2017	$54.94	$0.757	$68.26	$3.27	-$0.24	$68.02	$4.99	$73.24	$5.89	$74.15	$1.27	$69.52	$3.86	$72.12
2018	$60.68	$0.786	$72.95	$3.32	-$0.24	$72.71	$4.99	$77.93	$5.89	$78.84	$1.27	$74.21	$3.86	$76.81
2019	$65.35	$0.813	$76.20	$3.44	-$0.24	$75.96	$4.99	$81.18	$5.89	$82.09	$1.27	$77.46	$3.86	$80.06
2020	$69.35	$0.831	$79.41	$3.34	-$0.24	$79.17	$4.99	$84.40	$5.89	$85.30	$1.27	$80.68	$3.86	$83.27
2021	$73.66	$0.850	$82.62	$3.43	-$0.24	$82.39	$4.99	$87.61	$5.89	$88.52	$1.27	$83.89	$3.86	$86.48
2022	$76.40	$0.850	$85.79	$3.48	-$0.24	$85.55	$4.99	$90.77	$5.89	$91.68	$1.27	$87.05	$3.86	$89.65
2023	$78.29	$0.850	$87.96	$3.53	-$0.24	$87.72	$4.99	$92.94	$5.89	$93.85	$1.27	$89.22	$3.86	$91.82
2024	$80.23	$0.850	$90.17	$3.59	-$0.24	$89.93	$4.99	$95.15	$5.89	$96.06	$1.27	$91.43	$3.86	$94.03
2025	$82.18	$0.850	$92.42	$3.62	-$0.24	$92.19	$4.99	$97.41	$5.89	$98.32	$1.27	$93.69	$3.86	$96.28
2026	$84.19	$0.850	$94.48	$3.88	-$0.24	$94.24	$4.99	$99.47	$5.89	$100.37	$1.27	$95.75	$3.86	$98.34
2027	$85.85	$0.850	$96.34	$3.96	-$0.24	$96.10	$4.99	$101.33	$5.89	$102.24	$1.27	$97.61	$3.86	$100.20
2028	$87.57	$0.850	$98.28	$4.03	-$0.24	$98.04	$4.99	$103.27	$5.89	$104.18	$1.27	$99.55	$3.86	$102.14
2029	$89.34	$0.850	$100.25	$4.12	-$0.24	$100.01	$4.99	$105.24	$5.89	$106.14	$1.27	$101.52	$3.86	$104.11
2030	$91.13	$0.850	$102.27	$4.19	-$0.24	$102.04	$4.99	$107.26	$5.89	$108.17	$1.27	$103.54	$3.86	$106.13
2031	$92.95	$0.850	$104.30	$4.29	-$0.24	$104.06	$4.99	$109.29	$5.89	$110.20	$1.27	$105.57	$3.86	$108.16

Geological Formation	Pool Analogy	Pool Dimensions Width (ft)	Length (ft)	Area (Acres)	OOIP fctr (Bbl/ac-ft)	OGIP fctr (Mcf/ac-ft)	Ave Net Pay (ft)	OOIP (MBbl)	OGIP (MMcf)
Devonian	Rodney	4,175	4,175	400	1,329		13.4	7,127	
Silurian A-1 Gas	Mosald	2,952	2,952	200		249	16.0		796
Silurian Guelph Oil	Wanstead	2,087	2,087	100	443		35.0	1,551	
Silurian Guelph Gas	Ashfield	1,807	1,807	75		199	120.0		1,794
Silurian Clinton Gas	Norfolk	4,175	4,175	400		222	20.0		1,776
Ordovician Gas	Innerkip NP	1,500	15,000	517		428	15.0		3,313
Ordovician Oil	Goldsmith	1,500	15,000	517	607		42.0	7,379	
Cambrian Gas	N/A	7,000	7,000	1,125		413	20.0		9,288
Cambrian Oil	Clearville	7,000	7,000	1,125	414		20.0	12,149	

Production/Reserves		Prod wells (acre/well)	Inj wells (acre/well)	Initial Prod Rate per well (bpd/w) (Mcfd/w)	pool (bpd) (Mcfd)	Decline Rate (%/yr)	Rec Factor (%)	Recoverable Res (MBbl)	(MMcf)
Devonian	Vert WF	12.5	12.5	8	256	3.2%	35.4%	2,523	
	Hor WF	25	12.5	25	400	5.1%	36.0%	2,566	
Silurian A-1 Gas	Vert	50		100	200	7.7%	70.0%		557
	Hor	100		200	400	19.5%	75.0%		597
Silurian Guelph Oil	Vert WF	50	50	60	120	17.8%	15.0%	233	
	Hor WF	50	50	150	300	19.6%	35.0%	543	
Silurian Guelph Gas	Vert	75		300	300	5.5%	80.0%		1435
	Hor	75		750	750	17.0%	80.0%		1435
Silurian Clinton Gas	Vert	100		60	240	6.3%	50.0%		888
	Hor	200		120	240	6.3%	50.0%		888
Ordovician Gas	Vert	100		400	2000	30.1%	70.0%		2,319
	Hor	200		600	1800	25.2%	75.0%		2,485
Ordovician Oil	Vert	50		70	980	18.1%	26.0%	1,919	
	Hor	100		200	1400	19.9%	34.0%	2,509	
Cambrian Gas	Vert	100		150	1650	8.8%	70.0%		6,501
	Hor	400		600	1800	9.0%	75.0%		6,966
Cambrian Oil	Vert GF	100	500	40	400	9.3%	12.0%	1,458	
	Hor GF	150	500	100	600	8.6%	20.0%	2,430	

Economic Indicators		Wells	Capital Costs Fac	Ttl	Pool Op Costs Fixed	Variable	Net Present Value NPV12.5	NPV12.5	ROR
		($k)	($k)	($k)	($k/mo)	($/Bbl,$/Mcf)	($k)	($k)	(%)
Devonian	Vert WF	$7,619	$500	$8,119	$50	$10	$34,792	$26,497	53%
	Hor WF	$10,020	$500	$10,520	$60	$10	$39,267	$31,063	62%
Silurian A-1 Gas	Vert	$540	$400	$940	$10	$0.25	-$432	-$444	0%
	Hor	$1,180	$400	$1,580	$10	$0.25	-$618	-$636	0%
Silurain Guelph Oil	Vert WF	$1,380	$500	$1,880	$10	$10	$4,550	$3,971	87%
	Hor WF	$1,920	$500	$2,420	$15	$5	$14,763	$13,142	204%
Silurian Guelph Gas	Vert	$270	$400	$670	$10	$0.25	$867	$667	37%
	Hor	$590	$400	$990	$10	$0.25	$1,445	$1,267	68%
Silurian Clinton Gas	Vert	$1,280	$400	$1,680	$10	$0.50	-$818	-$857	0%
	Hor	$1,280	$400	$1,680	$10	$0.50	-$818	-$857	0%
Ordovician Gas	Vert	$3,150	$400	$3,550	$10	$0.50	$1,332	$1,086	32%
	Hor	$3,720	$400	$4,120	$10	$0.50	$892	$635	23%
Ordovician Oil	Vert	$9,450	$900	$10,350	$50	$8	$55,089	$48,731	173%
	Hor	$9,100	$900	$10,000	$50	$8	$78,404	$69,981	260%
Cambrian Gas	Vert	$6,930	$400	$7,330	$10	$0.25	$3,230	$2,050	22%
	Hor	$3,720	$400	$4,120	$10	$0.25	$6,956	$5,599	48%
Cambrian Oil	Vert GF	$8,100	$700	$8,800	$50	$10	$26,930	$22,404	75%
	Hor GF	$9,150	$700	$9,850	$50	$10	$49,168	$41,101	106%

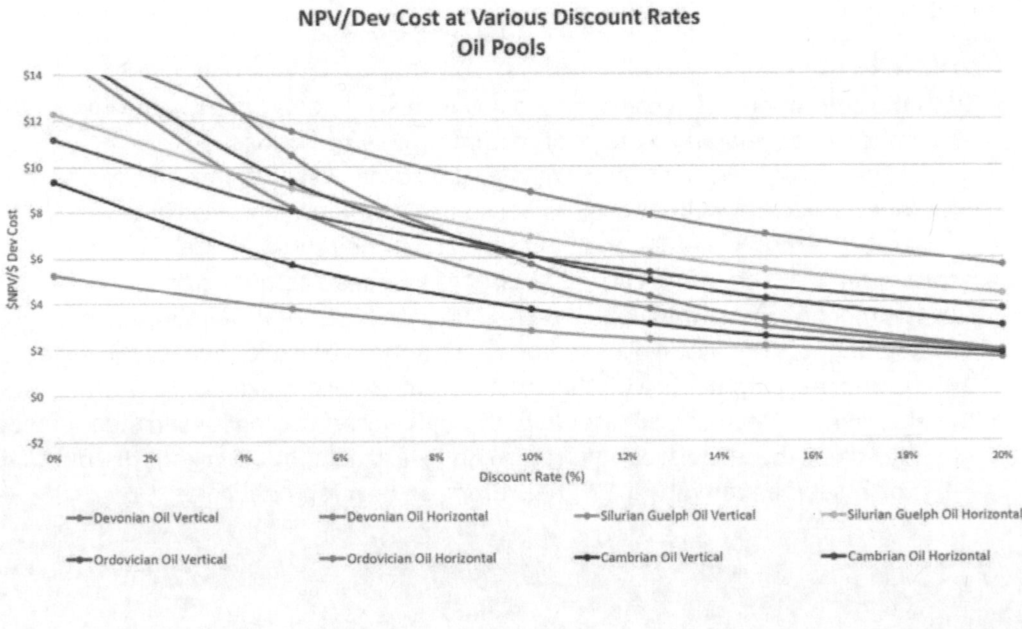

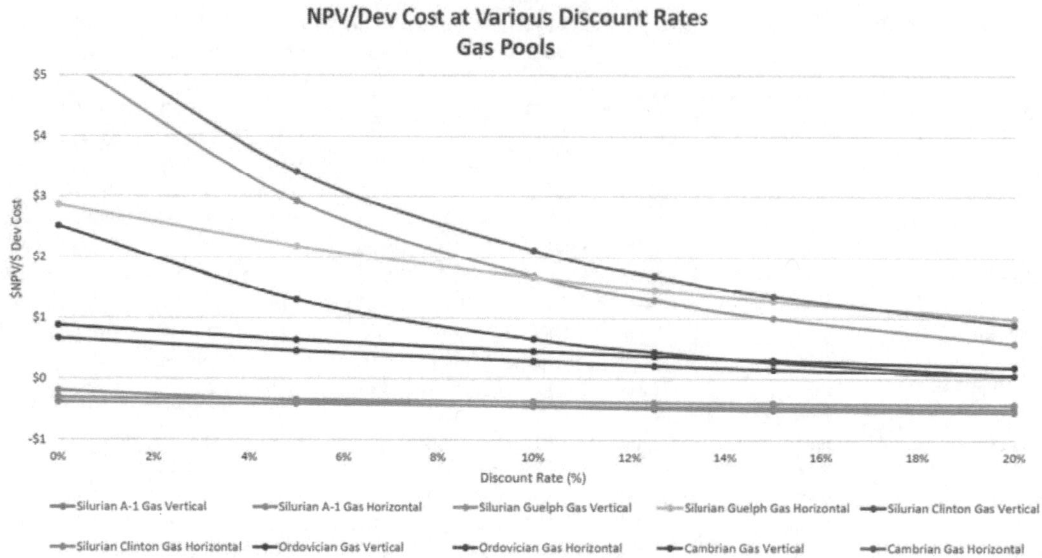

Conclusions

- With the current oil prices compared to gas prices, oil pools are much more economical to develop than gas pools within similar formations
- In most reservoirs, pool development with horizontal wells will improve the economics compared to vertical wells, mostly as a result of accelerated production
- Silurian A-1 Carbonate and Silurian Clinton sandstone reservoirs are likely not economical to develop with current gas prices and development costs
- Development of Silurian Guelph oil reservoirs with horizontal wellbores and development of Ordovician oil reservoirs with either vertical or horizontal wellbores are the pool developments with the most favourable economics
- Development of Devonian oil reservoirs through waterflooding for secondary recovery provides the highest undiscounted return on investment, but at higher discount rates other oil reservoirs can return a higher return on development costs

OIL AND GAS PLAYS OF THE MICHIGAN BASIN, SOUTHERN ONTARIO

PLAY BY PLAY
Exploring the Michigan Basin

Terry Carter, Consulting Geologist

London, Ontario

1

Ontario Petroleum Industries

- **Oil and Natural Gas**
  - 1250 oil wells, 1200 gas wells, annual production 400,000 bo, 5.7 bcf natural gas, 100 different producers
  - 5 wells/yr, 27,000 well records
- **Hydrocarbon Storage in Geological Formations**
  - 270 bcf natural gas in 35 depleted reservoirs, 275 wells
  - 22 million bbl refined petroleum products in 71 solution-mined caverns at Sarnia-Windsor area refineries/petrochemical plants – 95 wells
- **Salt Solution Mining**
  - 250,000 tonnes/yr, 18 wells

PLAY BY PLAY
EXPLORING THE MICHIGAN BASIN
2

Industry Participants

- **Oil & Gas**
  - Historically and presently exploration and production is dominated by small (mostly), Ontario-based operators
  - History of periodic interest from large Calgary-based and international companies
  - Small companies are low-cost, maintain operations through down-cycles, generate new plays, raise local capital
  - Local companies have grown into large national and international corporations with long-term economic impact;
    - Imperial Oil, Union Gas, and former McColl-Frontenac (Texaco Canada), British-American Oil Co. (Gulf Canada) and White Rose (purchased by Shell Canada)
- **Hydrocarbon Storage**
  - Natural gas storage dominated by one large +billion$ company
  - All cavern storage operations owned by large +billion$ petrochemical companies
- **Salt solution mining**
  - Two operations owned by large corporations

PLAY BY PLAY
EXPLORING THE MICHIGAN BASIN
3

Ontario Oil and Gas History & Firsts

- 1858 - first commercial oil well in North America at Oil Springs (and first oil spills)
- 1866 – first salt solution mining well at Goderich
- 1870 – first oil exports to U.S
- 1873 – first export of technology – first Canadian drillers leave for Indonesia
- 1889 - commercial gas production at Kingsville and Welland
- 1890 – first export of natural gas to U.S.
- 1913 – first offshore well in Lake Erie
- 1915 – first subsurface injection of natural gas for storage
- 1914 – Ontario Natural Gas and Oil Wells Act
- 1985 - year of peak gas production
- 1995 - year of peak oil production

PLAY BY PLAY
EXPLORING THE MICHIGAN BASIN
4

Geological Setting

-Precambrian crystalline rocks of Canadian Shield form core of the North American continent, > 1 billion years old

-Sedimentary rocks deposited on top of these crystalline rocks around edges of the continent

-Oil and gas production only in Michigan & Appalachian basins

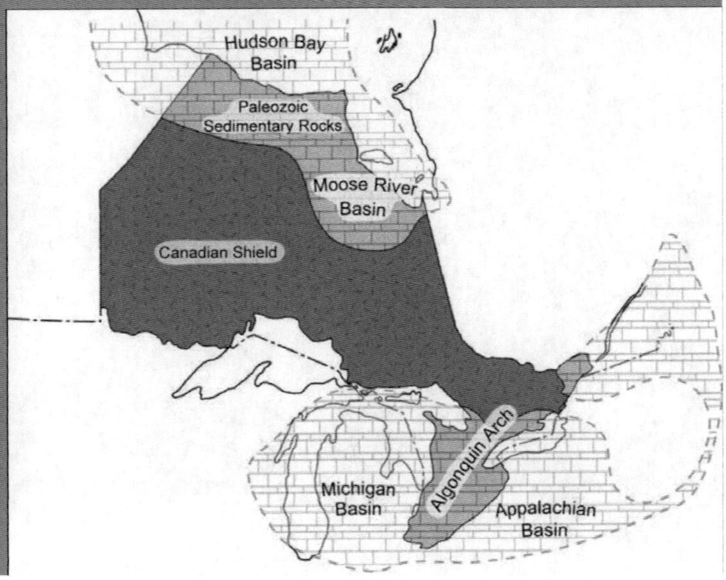

Bedrock Geology of Southern Ontario

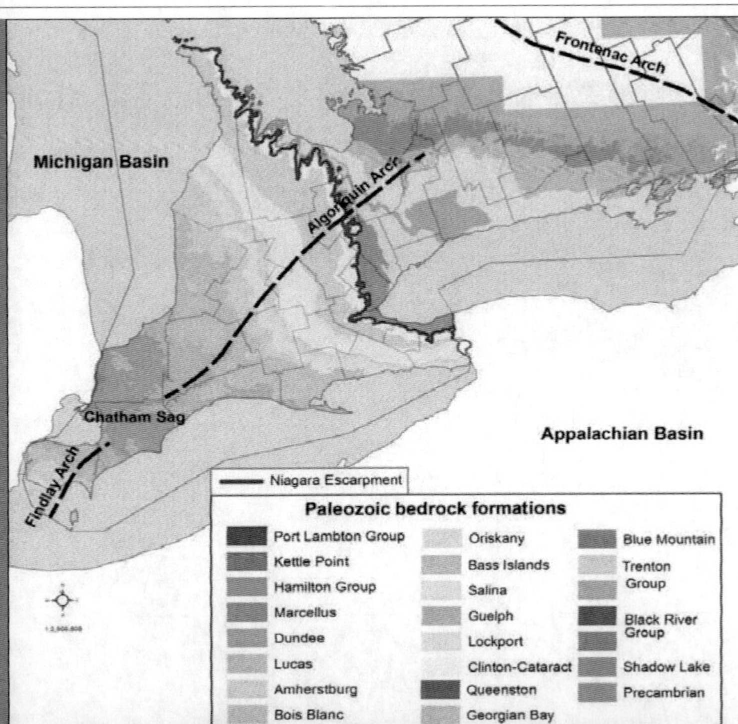

PLAY BY PLAY
EXPLORING THE MICHIGAN BASIN
6

Paleozoic Stratigraphy of Southern Ontario

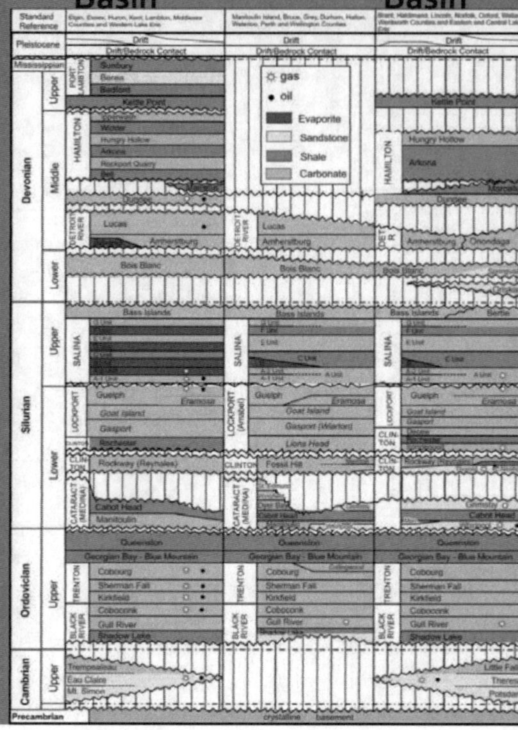

Oil and Natural Gas Fields of the Michigan and Appalachian Basins

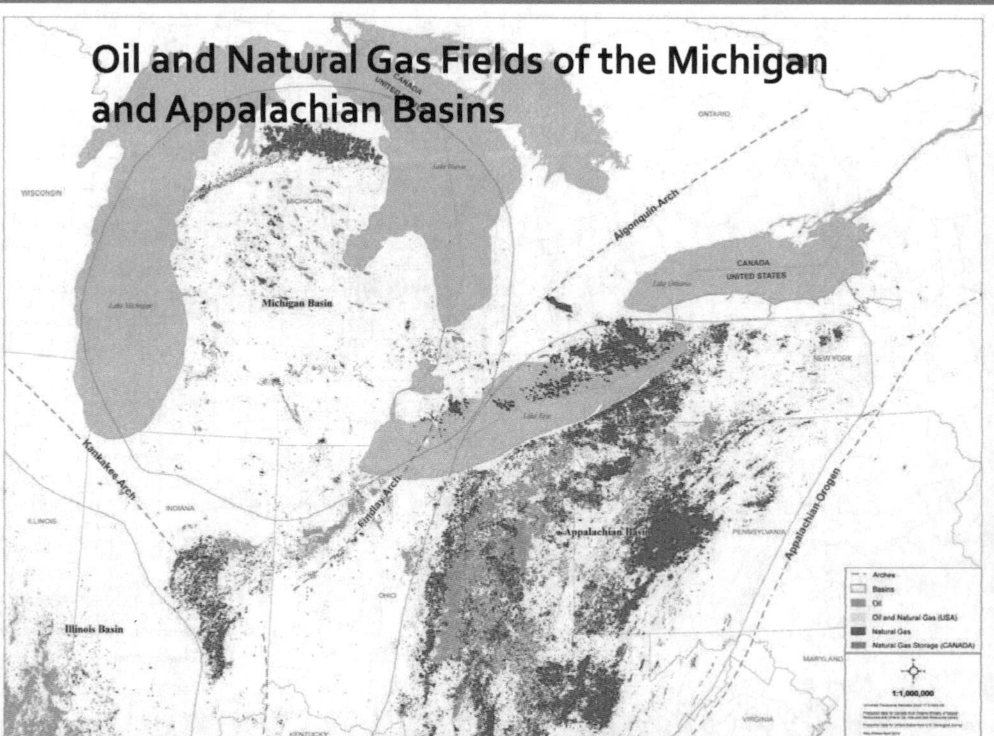

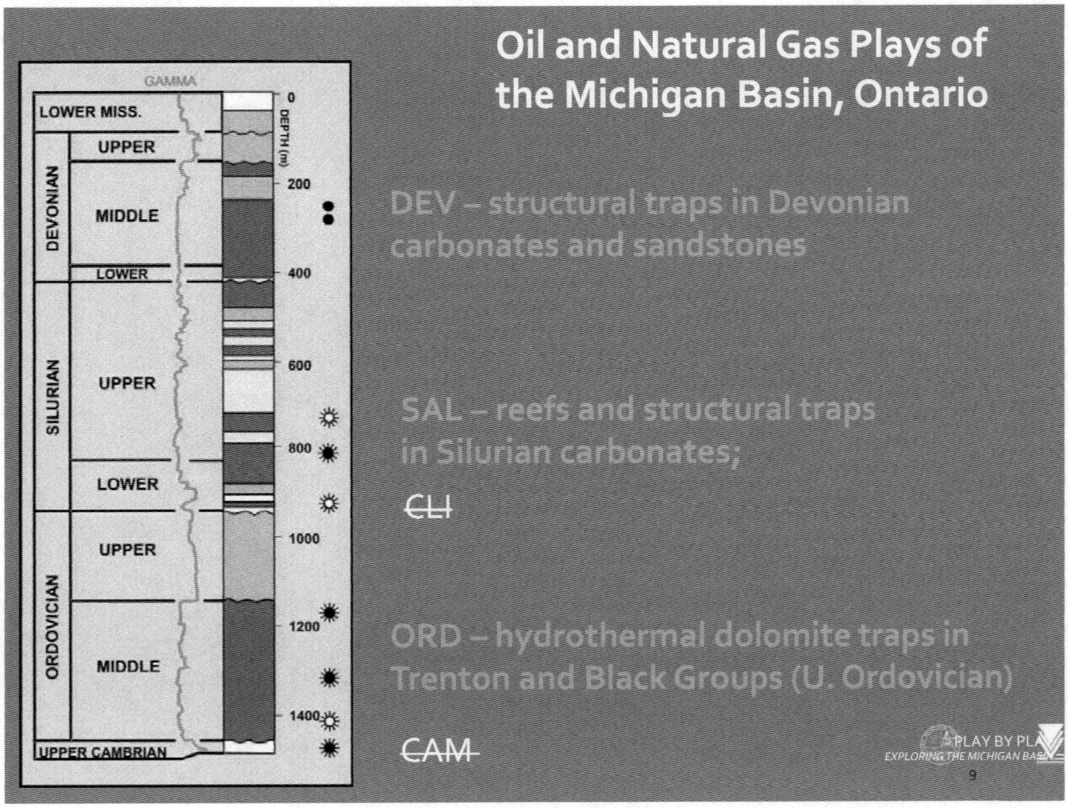
Oil and Natural Gas Plays of the Michigan Basin, Ontario
GAMMA
LOWER MISS.
DEVONIAN
UPPER
MIDDLE
LOWER
SILURIAN
UPPER
LOWER
ORDOVICIAN
UPPER
MIDDLE
UPPER CAMBRIAN
DEPTH (m)
0
200
400
600
800
1000
1200
1400
DEV – structural traps in Devonian carbonates and sandstones
SAL – reefs and structural traps in Silurian carbonates;
CLI
ORD – hydrothermal dolomite traps in Trenton and Black Groups (U. Ordovician)
CAM
PLAY BY PLA
EXPLORING THE MICHIGAN BAS
9

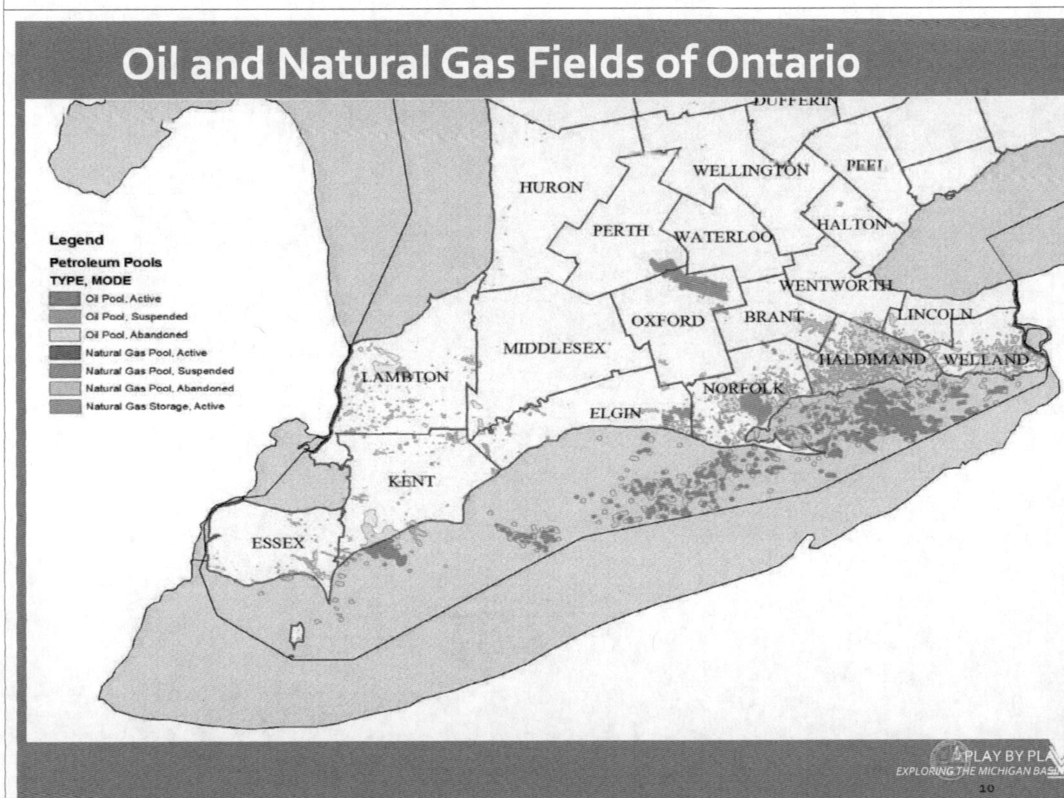
Oil and Natural Gas Fields of Ontario
Legend
Petroleum Pools
TYPE, MODE
Oil Pool, Active
Oil Pool, Suspended
Oil Pool, Abandoned
Natural Gas Pool, Active
Natural Gas Pool, Suspended
Natural Gas Pool, Abandoned
Natural Gas Storage, Active
DUFFERIN
HURON
WELLINGTON
PEEL
PERTH
WATERLOO
HALTON
WENTWORTH
OXFORD
BRANT
LINCOLN
MIDDLESEX
HALDIMAND
WELLAND
LAMBTON
NORFOLK
ELGIN
KENT
ESSEX
PLAY BY PLA
EXPLORING THE MICHIGAN BAS
10

Oil and Natural Gas Fields of the Michigan Basin, Ontario

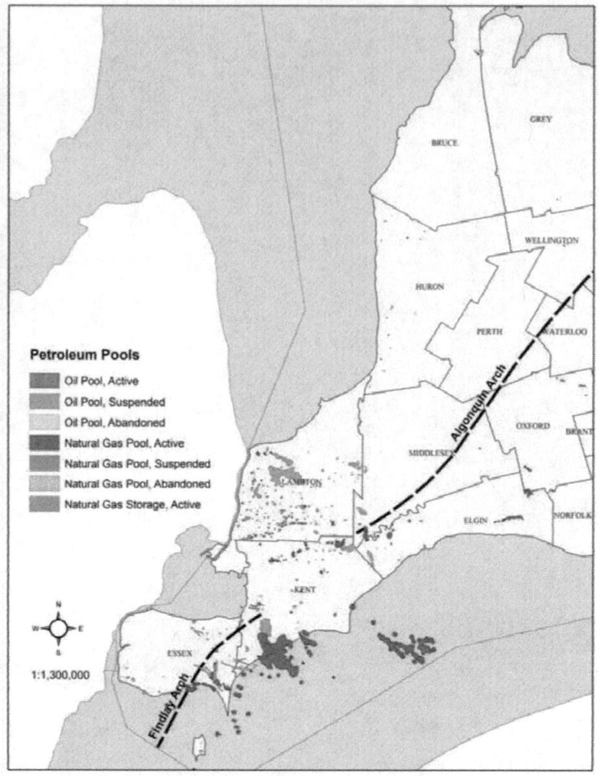

Ontario Oil and Gas Plays

Play	Description	Depth m.	Cum. production
DEV	Structural traps in fractured Devonian carbonates and sandstone – structural domes related to differential salt dissolution	100 - 150	**45 mmbo**
SAL	Silurian carbonates - pinnacle, incipient and patch reefs - fault traps	300 - 700	**15 mmbo** **748 bcf** **250 bcf gas storage capacity**
CLI	Lower Silurian sandstones basin-centre stratigraphic trap	150 - 500	0.05 mmbo 507 bcf
ORD	Ordovician carbonates – hydrothermal dolomite reservoirs	800 - 900	**25 mmbo** **43 bcf**
CAM	Cambrian sandstones – stratigraphic traps, fault traps	700 - 1200	5.3 mmbo 32 bcf

EXPLORING THE MICHIGAN BASIN

12

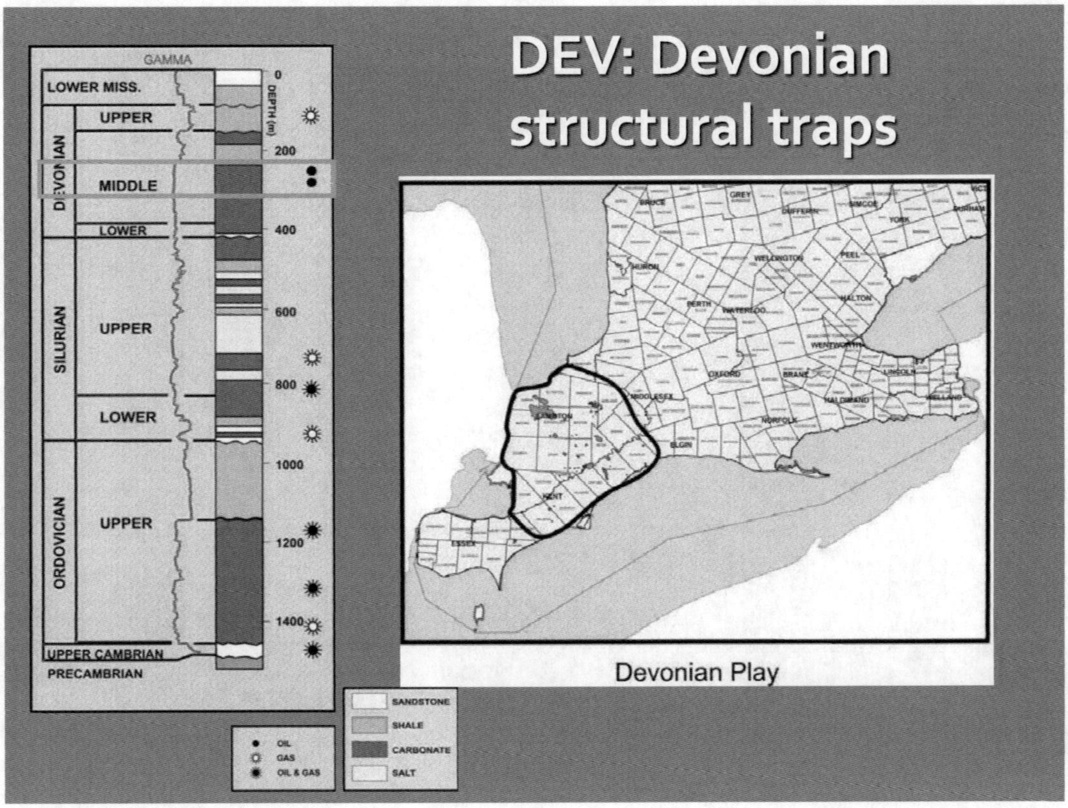
DEV: Devonian structural traps
GAMMA
LOWER MISS.
DEVONIAN
UPPER
MIDDLE
LOWER
SILURIAN
UPPER
LOWER
ORDOVICIAN
UPPER
UPPER CAMBRIAN
PRECAMBRIAN
DEPTH (m)
0
200
400
600
800
1000
1200
1400
Devonian Play
OIL
GAS
OIL & GAS
SANDSTONE
SHALE
CARBONATE
SALT

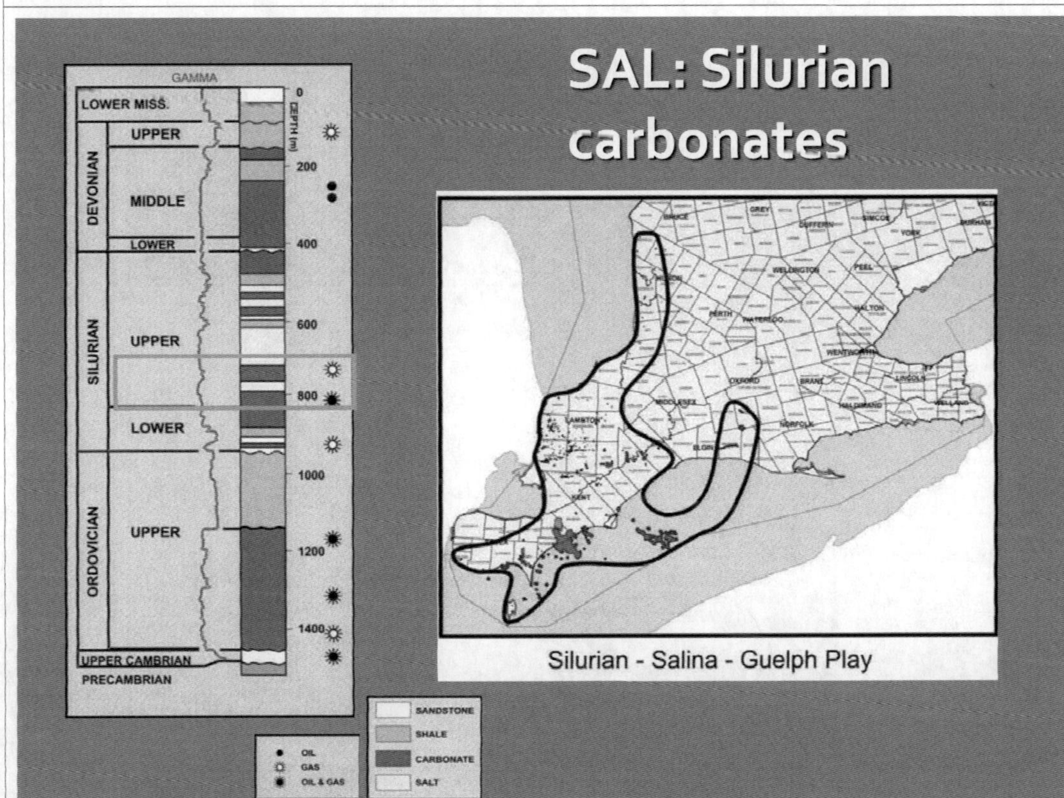
SAL: Silurian carbonates
GAMMA
LOWER MISS.
DEVONIAN
UPPER
MIDDLE
LOWER
SILURIAN
UPPER
LOWER
ORDOVICIAN
UPPER
UPPER CAMBRIAN
PRECAMBRIAN
DEPTH (m)
0
200
400
600
800
1000
1200
1400
Silurian - Salina - Guelph Play
OIL
GAS
OIL & GAS
SANDSTONE
SHALE
CARBONATE
SALT

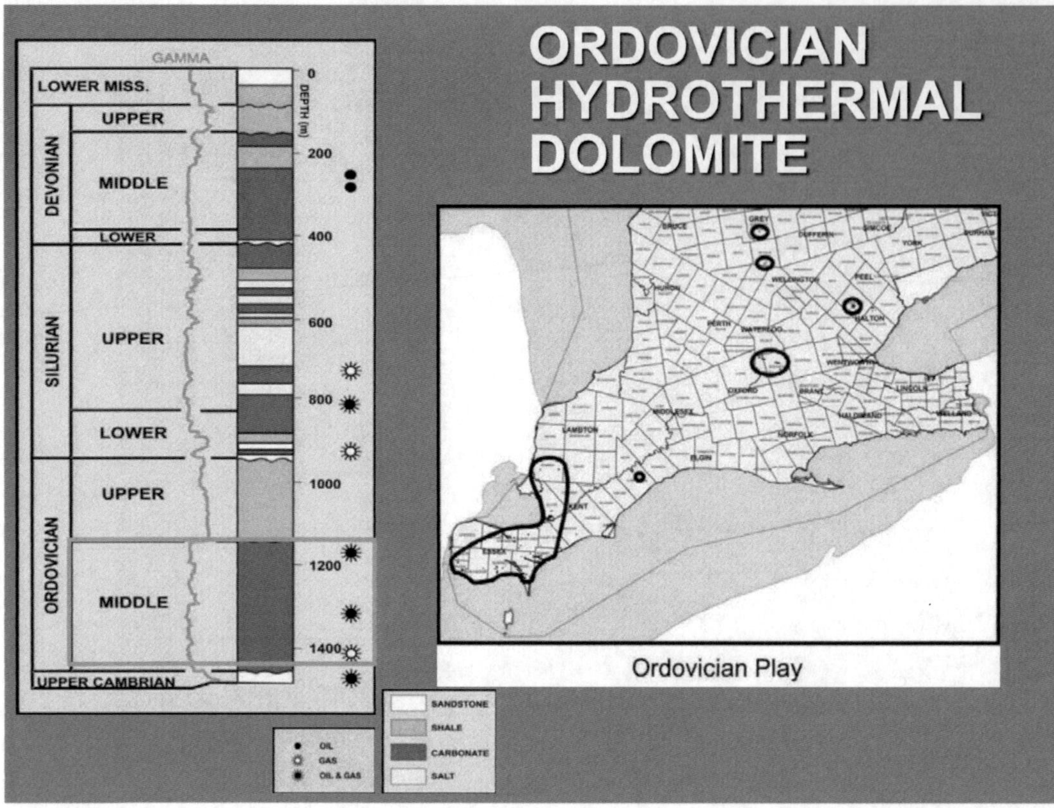
ORDOVICIAN HYDROTHERMAL DOLOMITE
GAMMA
LOWER MISS.
DEVONIAN
UPPER
MIDDLE
LOWER
SILURIAN
UPPER
LOWER
ORDOVICIAN
UPPER
MIDDLE
UPPER CAMBRIAN
DEPTH (m)
0
200
400
600
800
1000
1200
1400
Ordovician Play
OIL
GAS
OIL & GAS
SANDSTONE
SHALE
CARBONATE
SALT

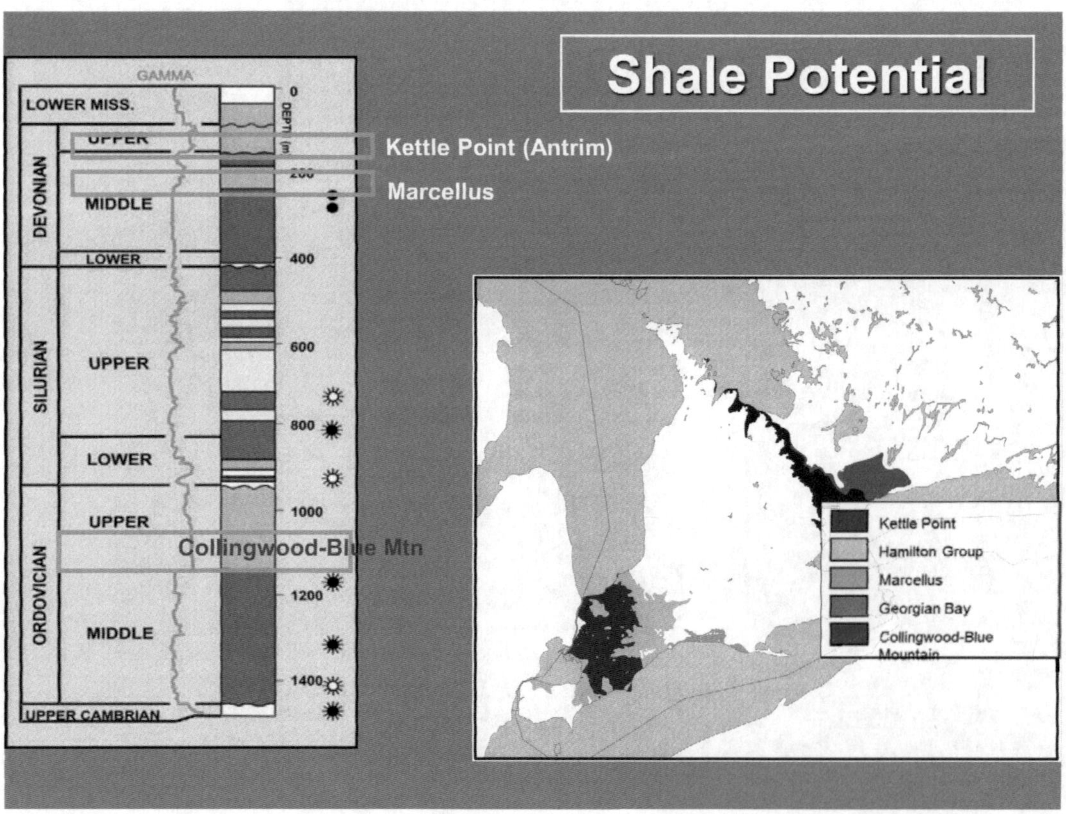
Shale Potential
GAMMA
LOWER MISS.
DEVONIAN
UPPER
MIDDLE
LOWER
SILURIAN
UPPER
LOWER
ORDOVICIAN
UPPER
MIDDLE
UPPER CAMBRIAN
DEPTH (m)
0
200
400
600
800
1000
1200
1400
Kettle Point (Antrim)
Marcellus
Collingwood-Blue Mtn
Kettle Point
Hamilton Group
Marcellus
Georgian Bay
Collingwood-Blue Mountain

POTENTIAL SHALE PLAYS

	Kettle Point	Marcellus	Collingwood-Blue Mountain
Thickness	30-105	1-12	10-50
Max depth m.	143	225	1000
Area km^3	9500	4700	70,000
TOC	3 – 15%	1 – 11%	1 – 11%

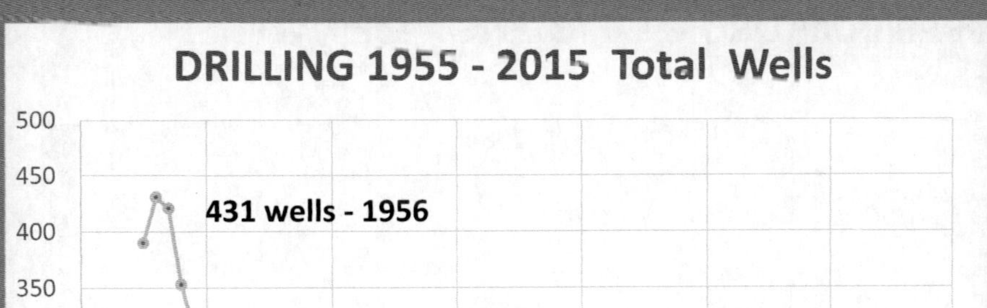

PLAY BY PLAY
EXPLORING THE MICHIGAN BASIN
18

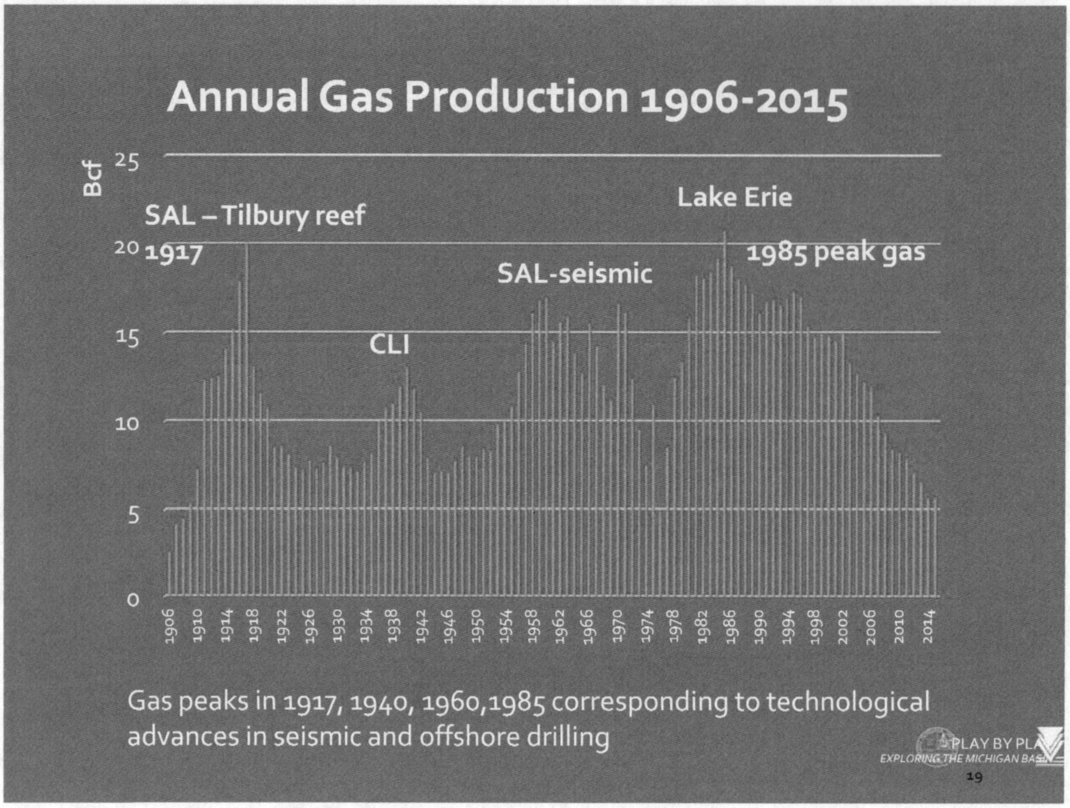
Annual Gas Production 1906-2015
Bcf
25
20
15
10
5
0
SAL – Tilbury reef
1917
CLI
SAL-seismic
Lake Erie
1985 peak gas
1906 1910 1914 1918 1922 1926 1930 1934 1938 1942 1946 1950 1954 1958 1962 1966 1970 1974 1978 1982 1986 1990 1994 1998 2002 2006 2010 2014
Gas peaks in 1917, 1940, 1960,1985 corresponding to technological advances in seismic and offshore drilling
PLAY BY PLAY
EXPLORING THE MICHIGAN BASIN
19

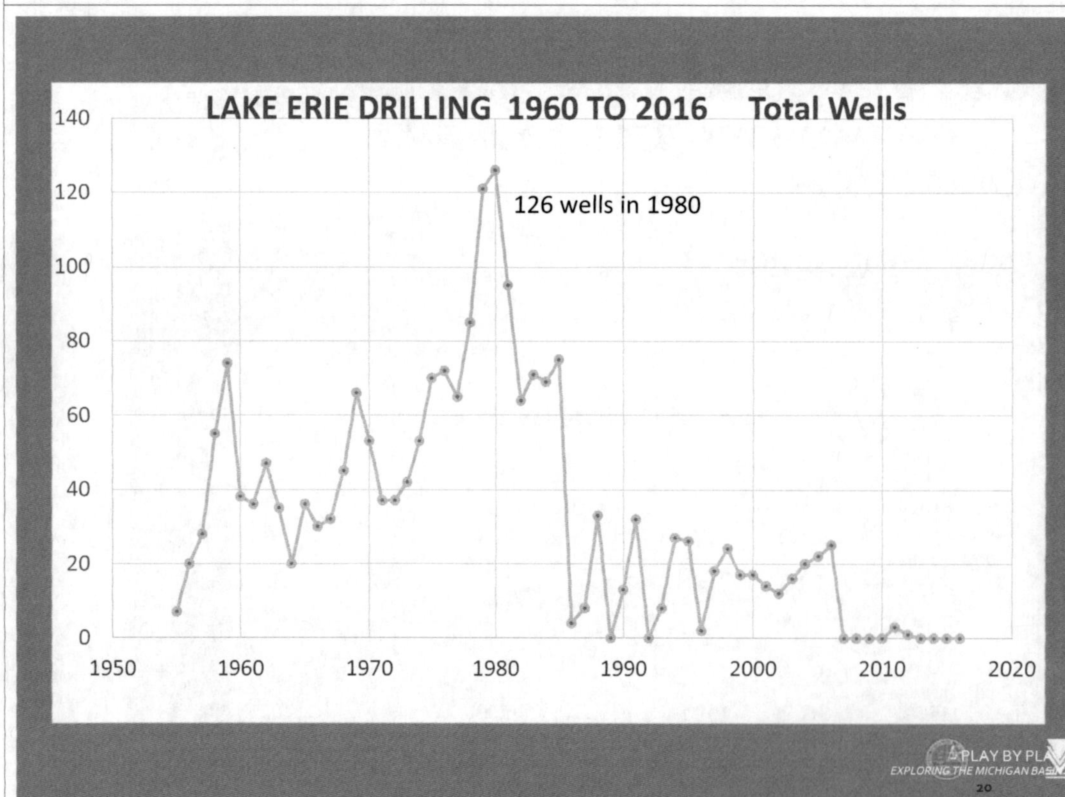
LAKE ERIE DRILLING 1960 TO 2016 Total Wells
140
120
100
80
60
40
20
0
126 wells in 1980
1950 1960 1970 1980 1990 2000 2010 2020
PLAY BY PLAY
EXPLORING THE MICHIGAN BASIN
20

Annual Oil Production 1863 - 2015

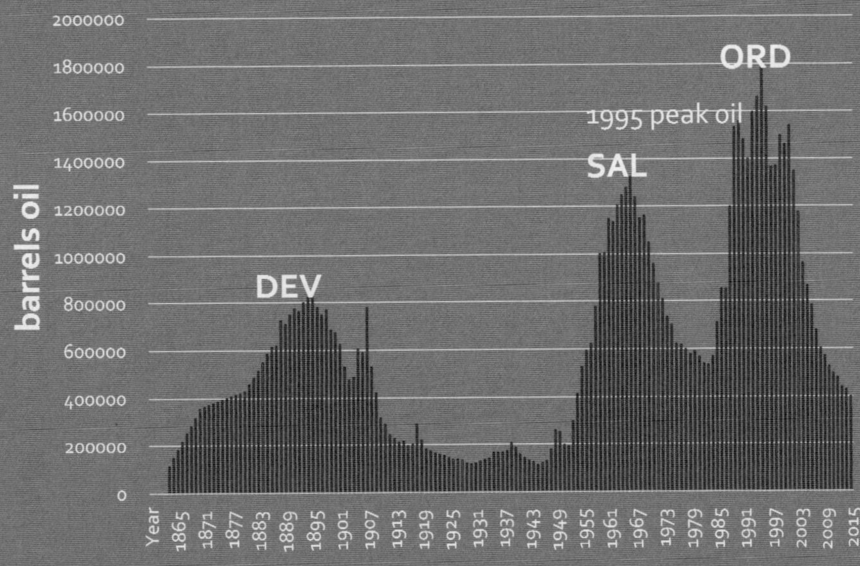

Oil peaks in 1895, 1966, and 1995 corresponding to successive discovery of deeper plays, seismic and new exploration models

21

Oil, Gas and Salt Resources Library

- Not-for-profit centre for management of publicly accessible data on oil, gas, salt, hydrocarbon storage resources, and subsurface Paleozoic geology of Ontario
- Industry-operated
- Both free and fee-based access to data
- 669 Exeter Road, London, ON
- www.ogsrlibrary.com

Data Resources

- Petroleum well files 26,500 +20/yr
- Scanned well file images 500,000+
- Drill cuttings 11,000 wells +20/yr
- Drill core 1,100 +4/yr
- Monthly production reports 40,000 +2,000/yr
- Injection+disposal reports 10,000 +150/yr
- Maps & Reports 2,500 +10/yr
- Geophysical logs 21,000
- Oil and gas pool map
- Journals, reprints, government reports
- ArcGIS workstation
- Digital petroleum well database
- Oil, gas, water analyses, isotopic analyses
- Deep groundwater maps

  89 water type, 17 potentiometric

23

Digital Data: www.ogsrlibrary.com

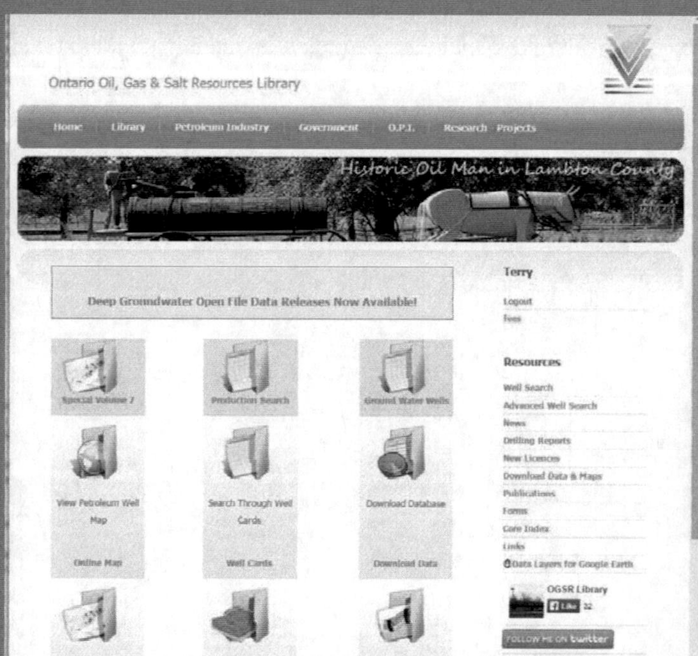

What Next?

- Large unexplored areas in Ordovician
- Undiscovered incipient reefs
- Low drilling density in Huron County
- Lake Huron reefs?
- Conversion of more depleted gas pools to storage
- Technological improvements in enhanced recovery from known oil reservoirs
- Historic Devonian oil pools now account for 20% of Ontario annual production – flat production decline
- New play concepts
- Unassessed shale gas/oil potential
- Excellent data availability at OGSR Library

PLAY BY PLAY
EXPLORING THE MICHIGAN BASIN
25

QUESTIONS?

PLAY BY PLAY
Exploring the Michigan Basin

ONTARIO PETROLEUM INSTITUTE INC.
1858
FIRST COMMERCIAL WELL IN NORTH AMERICA
O.P.I.

26

AUTHOR INDEX

Ontario Petroleum Institute (OPI)
555 Southdale Road East
Suite 203
London, Ontario, Canada
N6E 1A2

ISBN 978-1-5108-4650-0

CURRAN ASSOCIATES INC.
proceedings.com